Farmers' Markets

Farmers' Markets

Success, Failure, and Management Ecology

Garry Stephenson

CAMBRIA
PRESS
Amherst, New York

Printed in the United States of America

Library of Congress Cataloging-in-Publication Data

Stephenson, Garry Owen.
Farmers' markets : success, failure, and management ecology / Garry Stephenson.
p. cm.
Includes bibliographical references and index.
ISBN 978-1-60497-509-3 (alk. paper)
1. Farmers' markets—United States. I. Title.

HF5472.U6S74 2008
381'.410973—dc22

2008001969

For Geoff, Landon, and Karissa

Table of Contents

List of Figures

List of Tables

Foreword

It is a rare treat to read through a book on farmers' markets and feel like you are eating a new fruit with a unique flavor, yet also with a satisfying familiarity that makes one feel nourished by the experience. When Garry Stephenson asked me to write a foreword to his new book on farmers' markets, I was surprised and honored. Because he is one of the prominent applied researchers on farmers' markets in the United States, I also was humbled. I am delighted to contribute to a work which is a wonderfully practical, yet thoroughly researched, overview of the life of farmers' markets.

The first decade of the 21st century has witnessed another "renaissance" of farmers' markets in the United States. Researchers and journalists alike have documented the flowering of this community institution with its attendant social, economic, and nutritional benefits for consumers and local farmers (Gillespie, Hilchey, Hinrichs, & Feestra, 2007). Recently, the public's desire for more "local" food, however that is defined, has been a boon for farmers' markets across the country. We read accounts and articles about how farmers' markets positively impact the environment, towns, neighborhoods, and the lives of those who participate. What Stephenson's book adds to these many and varied snapshots of particular markets or points in time is a historical context and some very interesting lessons about market dynamics as seen from the eyes of an anthropologist and the director of a small farm program in a land grant university.

Stephenson manages to walk a very elusive line between the theoretical requirements and interests of the academic community on the one hand, and the practical needs and interests of farmers' market participants and practitioners on the other. In his straight-forward language, he starts with a colorful account of farmers' markets from the turn of the last century to today, adding insights about what factors have contributed to the unprecedented growth we have seen recently. He looks in more depth, from a variety of perspectives, at the many particular farmers' markets in Oregon, his home state, shedding light on how they work best and what challenges they face from the very local operational level to the national policy level.

For me, one of the unique elements of the book was the new understanding I gained from Stephenson's practical, problem-based analysis of farmers' markets, using political ecology as an analytical framework. This approach allows him to look at farmers' markets from the contexts of both their natural and political environments, and to see the interactions between these two. In particular, Stephenson applies this political-ecology framework to the management of markets, which opens up new ways to understand the years of data and experience he has gathered about market successes and failures in Oregon. Market managers have already found the visual tool in Chapter 6 he created helpful in their own market assessments. Instead of a "how-to" recipe approach, Stephenson's framework and visual tool allow both market managers and researchers to gain a more comprehensive view of the market(s) in which they are interested. One senses that if this tool were taken seriously, market managers, researchers, and interested members of the public could discover many new dimensions of their markets and find ways to improve them.

One does not have to read far before realizing that Stephenson's writing is grounded in many years of experience with farmers' markets. He visited dozens of farmers' markets in Oregon and throughout the West. For me, the familiar flavor of this book comes from the pragmatic analyses of market scale, market management, and community interactions, and how these impact market successes and failures. With all the focus on successful markets in the media, we often do not hear about why some markets fail. Stephenson's chapters about market failures are as instructive as those about what makes markets succeed. Although there are many individual cases, Stephenson is able to reveal some interesting patterns and themes.

Stephenson has taken great pains to include a diverse advisory committee in the design and analysis of his research, both for the new ideas they offer and to make sure his conclusions are grounded in reality. In this participatory methodology, market participants and researchers worked together to synthesize findings and discuss new ideas that emerged. This participatory framework is woven throughout the book and contributes to the pragmatic, realistic themes and conclusions.

The work behind this book represents some of the best public scholarship as defined by Scott Peters (Peters, Jordan, Adamek, & Alter, 2005). Stephenson's practical, participatory, research-based work on farmers' markets fits Scott Peters' (2005) description of public scholarship perfectly—"creative, intellectual work that is conducted in public, with and for particular groups of citizens. Its results are communicated to, and validated by, peers, including but not limited to peers in scholars' academic fields. Scholars...advance the academy's teaching and research mission in ways that hold both academic and public value" (p. 14).

As Peters says, one of the principal differences between public scholars and public intellectuals is that the former work *with* various publics and the latter work *on behalf of* the publics they study. The notion of a "public scholar" complements one of the tenets of a sustainable food system—that community residents participate actively in its design and implementation (UC SAREP, n.d.). Farmers' markets are indeed one of the tangible hallmarks of a sustainable food system in many communities.

Finally, Stephenson adds a sparkle to his scholarship through the vibrant descriptions of activities that take place at markets. Writing as an anthropologist would record all the details of daily life in a community, Garry shares all the sights, sounds, and smells of several markets. You can almost feel and taste the tart apples being offered to a customer, or feel the cold, dripping rain run off market canopies onto the head of a vendor. He describes shoppers, young and old—some bargain-hunting, others mainly enjoying a cup of warm coffee and the stroll through the market, others still trying to navigate the market with strollers or with arms loaded with lugs of fresh strawberries. These descriptions remind the reader that these markets are not theoretical constructs or units of analysis, but live, busy places that are part of community life.

In the end, farmers' markets are about community relationships—between farmers and consumers, managers and vendors, government agencies and managers, or community organizations/businesses and market boards. These relationships are complex and dynamic. They are both what make farmers' markets so inviting and wonderful, and what burn some managers (and markets) out. But, if I were a betting woman, I'd say farmers' markets are here to stay. Stephenson's contribution with this book allows us, whether we are managers, researchers, farmers, or eaters, to understand how all community members can work in concert to make this institution a healthy and vibrant part of a sustainable food system.

Gail Feenstra
Food Systems Analyst
Sustainable Agriculture Research & Education Program/
Agricultural Sustainability Institute
University of California, Davis

ACKNOWLEDGMENTS

I am grateful to those who reviewed this document in its previous forms. Geraldine Moreno-Black, Phil Young, Stephen Wooten, Daniel Pope, and Court Smith were invaluable for their critical reading and evaluation. My sincere thanks go to my colleagues, Larry Lev and Linda Brewer, for their substantive reviews of portions of this document, and again to Linda Brewer for her role as research assistant in segments of this project.

The Project Advisory Committee (Dianne Stefani-Ruff, Suzanne Briggs, Rebecca Landis, and Valerie Blaha), the farmers' market managers, members of farmers' market boards of directors, and participating farmers who generously gave their time to assist with this research project have my enduring gratitude.

This investigation was supported, in part, by the United States Department of Agriculture through funding from the Initiative for the Future of Agriculture and Food Systems. Additional support was provided by the Extension Agriculture Program of the College of Agricultural Sciences, Oregon State University.

Farmers' Markets

INTRODUCTION

* * * * *

Saturday, August 20. The peaches are the size of my fist. Only August can do that. It's 8:35 a.m., and as I walk toward the market, there are people already walking away with bags of produce. The market doesn't officially open until 9:00. It's a perfect morning.

In the market, a customer slowly strolls. An empty canvas bag swings from one hand, coffee in the other. Kids wave sunflowers like wands. There is the occasional whiff of basil. A mountain of sweet corn fills the backs of two pickup trucks parked side by side—*Four ears for $1*. A babyless stroller is filled with flowers and greens; dad offers a piece of scone to the toddler balancing alongside.

There are about 43 vendors today. Some vendors require multiple spaces, enlarging the scale of the market. There are vegetables, fruit, cut flowers, nursery plants, bakery products, fresh fish, cheeses, meats, nuts, mushrooms, honey, and more. The market stretches to about 600 feet long, down both sides of 1st Street. Gates at either end of the market close the street. The site is sufficiently large to have musicians at each end out of earshot from each other: acoustic blues at one end, bluegrass at the other.

A woman with a net market bag drags carrot tops on the pavement as she walks. Another leans forward as she bites into a peach to let the juices fall on the street.

(*continued in the "August Vignette" on p. 43*)

* * * * *

This sun-soaked farmers' market at the height of its season is part of a phenomenon. Farmers' markets in Oregon and across the United States are experiencing growth in numbers and exceptional popularity with consumers. More than 70 farmers' markets operated in Oregon during 2006, a result of steady growth over recent decades. Most of this growth is recent, with more than 50 new farmers' markets opening in the past 10 years. Nationally, farmers' markets have increased at a similar pace numbering over 4,385 as of 2006, an addition of over 2,600 markets in 12 years.

The reasons for this growth are complex. While there is a trend toward an increasing consolidation and globalization of agriculture and commoditization of food, there is also opposition to this trend through an emergence of interest in locally produced food for health and good living, and as a component of communities and their local economies. Farmers' markets are a crucial marketing channel for small farmers who, as a group, have been excluded from traditional wholesale markets because the extremely low profit margins associated with those markets require a production level beyond the scale of a small farm. Small farmers have used farmers' markets and other direct marketing channels to create a new entrepreneurial agriculture that operates independently and provides a gateway into farming for those with limited resources.

In Oregon, the farmers' markets many of these small farmers depend on were conceived and initiated and grew independently of one another in individual communities. The markets are funded almost entirely by farmers' fees. They are community-based grassroots organizations that operate on "shoestring" budgets with little in the way of resources for customer research, strategic planning, or the other practices that businesses and organizations access in order to ensure their longevity. Perhaps as a result, the growth in the number of farmers' markets has not been entirely smooth. Each year there are farmers' markets that are forced to close.

Between 1998 and 2005—a period for which there are reliable data on market start-ups and closings—Oregon's farmers' markets expanded in number from 38 to 68. While this is seemingly an impressive gain, it is distressing to note that during those eight market seasons twice as many markets actually opened, and more than 30 failed. Farmers' market organizers, customers, small farmers, and local food system/security advocates and practitioners are concerned about this issue and are interested in enhancing this important market channel and hub for communities. An overarching question is: *With farmers' markets increasing in popularity and numbers, why do some markets thrive while others do not?* The research findings presented here contribute to an answer.

The research for this book was conducted during a fascinating period marked by the interest of many consumers in locally produced foods, a resistance to a globalizing food system, and other trends. The growth in numbers of farmers' markets is a product of this period. Just as Jane Pyle (1971) conducted research during the late 1960s for her article "Farmers' Markets Functional Anachronisms" at the nadir of farmers' market popularity (unbeknownst to her), this research was conducted during a period of seemingly boundless interest in and support for farmers' markets.

This book has both academic segments and more practical segments. Students with interests in food-system research, along with practitioners managing or working with farmers' markets, should find portions of the book informative. A major goal of the book is to provide information and recommendations for market organizers and those who work with them, in order to assist with their decision making and strategic planning. The book explores historic trends related to growth and decline in market numbers, it examines the management organization associated with markets of specific sizes, it looks at the characteristics and issues associated with markets that fail, and it offers a model that illustrates how farmers' market organizers successfully adapt to barriers and challenges in their environments. Although the focus of this research is on one area in the United States, the findings have broad application.

The research has an applied orientation; that is, the findings provide the foundation for recommendations regarding market organization and planning that may enhance the success and longevity of individual farmers' markets. The research engages a "node" in the food system that has implications for the plight of small farmers and the social and economic life of communities. It focuses on a single thing that has possible implications for many things.

The methods utilized for this research included both quantitative and qualitative approaches. The approaches included a survey questionnaire, administered by telephone, of farmers' market managers, as well as interviews, participant observation, and focus groups. Throughout the text, research informants' identities have been withheld to protect anonymity. Quantitative and qualitative approaches were utilized in a manner that improved the quality of data collected and triangulated findings to enhance their validity, leading to an accurate picture of farmers' markets in Oregon.

As applied research, the data collection had a participatory approach that included a Project Advisory Committee composed of farmers' market managers and focus groups of research participants. The Project Advisory Committee helped formulate research questions, and they reviewed and critiqued the findings. The focus groups provided reaction to and synthesis of initial findings. The research approaches maintained a practical grounding and problem orientation to the research project. Details of research methods are provided in the Appendix.

OVERVIEW OF THE BOOK

Chapter 1, "ADM, a Tomato Named Local Lucy, and Small Farms: The Ecology and Reemergence of Farmers' Markets," reviews the literature related to this study. It utilizes literature related to broader themes, including discussions of the challenges small farmers face in a corporatized and globalizing agro-food system, the reemergence and function of local food systems, a review of ecological theory to put these processes in context and provide a framework for analysis, the need for concise applied research that works toward solving real-world problems, and a general summary of literature on farmers' markets.

Chapter 2, "The Nature of Oregon's Farmers' Markets," organizes basic demographic data on farmers' markets to show their intimate ties to the natural world. Farmers' markets connect people with the natural environment by means of their open-air adaptation and their policy of allowing only the sale of locally produced foods, thus fostering connections to the seasons and the individual agro-ecozones of Oregon.

Chapter 3, "Ninety Years of Oregon's Farmers' Markets: An Inventory of Markets and External Influences on Their Growth and Decline," uses a historic perspective to examine some of the major external influences on farmers' market growth and decline from 1915 to 2005. These influences have included economic downturns, wars, the extraordinary transformation of agriculture and the food system during the 1950s and 1960s, and the impact of political action and public policy during the 1970s. Although these influences are associated with periods of growth and decline in numbers of farmers' markets during the past, researchers have not adequately addressed what is driving the current rapid growth of markets since the early 1990s. This research indicates that one of the influences on the current growth in the number of farmers' markets is their use to build *community*.

Chapter 4, "Connecting Farmers' Market Size and Management Organization," examines the association between management organization and market scale. Market size categories are compared to the tools and resources commonly used for farmers' market management. There are significant differences in management tools used by markets of differing sizes. Findings indicate that in order to avoid internally inflicted problems that could negatively impact a market, organizers should use scale-appropriate management tools. In addition, to avoid outgrowing their management structure and to ensure the readiness of their market for future growth, organizers should plan for implementing appropriate management practices to match an anticipated market size.

Chapter 5, "When Things Don't Work: Some Insights into Why Farmers' Markets Close," reveals a dark side to the heady period of market growth we are experiencing. Although markets are growing in number, a very high number of them fail. In addition, farmers' markets experience significant rates of manager turnover. Underlying causes include a wide disparity in revenue to provide for market operations. This disparity impacts the ability of market organizers to hire labor to perform functions that help grow and sustain markets. Smaller markets often experience a *circular condition* in which they cannot attract sufficient customers because they do not have sufficient vendors, but they cannot attract sufficient vendors because they do not have sufficient customers.

Chapter 6, "It's Harder Than It Looks: The Agro-Social-Economic-Regulatory Ecology of Oregon's Farmers' Markets," utilizes elements from the previous chapters with additional data to create a synthesis of research findings. Individual farmers' markets appear and operate in the manner they do because of many minute adjustments made by market managers to uniquely adapt a market to influences from the natural and political environments. Manager-determined traits of successful markets are used, and an ecological framework explores the viable adaptations of farmers' markets to physical and political environmental influences. A visual model is constructed of the ecology within which farmers' markets function.

Chapter 7, "Concluding Observations and Recommendations," provides a general discussion of the research and some of the broad themes that emerged, summarizes the major research findings within an ecological theory framework, and offers recommendations to enhance the viability of individual farmers' markets and of farmers' markets collectively.

Lastly, for variety, two vignettes of a farmers' market offer a taste of a market in operation during different seasons. Presented in a stream-of-consciousness style, the vignettes serve as bookends to the data-laden chapters.

Chapter 1

ADM, a Tomato Named Local Lucy, and Small Farms: The Ecology and Reemergence of Farmers' Markets

Two Food Streams

Farmers' markets are at the center of a convergence of forces. On the one hand, there is an increasing commoditization, consolidation, and globalization of agriculture, with its negative consequences for farmers, consumers, and communities. On the other hand, there is an emergence of interest in food as a component of communities and associated local economies. This latter force is driven by a variety of issues including interests in and concerns about health and food, the re-development of cuisine based on locality, interests in food as a focal point for building community, strong concerns about the environmental consequences of industrial agriculture, and support for small farmers. Mark Grey (2000) described the situation this way: "The United States has experienced the development of two food 'streams.' One is based on an industrial model and is dominated by a handful of large, integrated transnational corporations. The other relies on direct marketing that links food producers with consumers. Although the latter is small by the standards of industrial agribusiness, it is growing rapidly" (p. 143).

The issue of two food streams is best framed by examining the advertising message of a transnational corporation and the advocacy message of an environmental organization. Archer Daniels Midland (ADM), in its widely distributed print and television advertising campaign from 2004, offers this vision of the future of agriculture:

> What if we looked at the world as one giant farm field?
>
> In tomorrow's global food economy, every crop will grow where it grows best.
>
> And, ADM can link farmers to almost any market in the world. It's a natural way to improve agricultural efficiency, make food more affordable, and feed a hungry world.
>
> Nature has answers. Is anyone listening?
>
> Yes.
>
> ADM.
>
> The nature of what's to come. (Archer Daniels Midland, 2004)

Here, ADM advocates a global food system and its own leading role in this system. This message is consistent with the views of Steven Blank (1998), an agricultural economist who argued that the United States has advanced technologically to a point where food production is no longer the best use for its land and capital. Noting that commodity prices are now set globally but production costs are set locally, he reasons that the United States no longer has a comparative advantage in food production because of its higher labor and land costs. So, for ADM, many of the farm fields it refers to are likely to be in developing countries, and the improved agricultural efficiency will come from land values and labor costs that are substantially lower than in the United States, Canada, or the European Union, and from an environmental and labor-regulatory atmosphere that is conducive to corporate profitability. ADM's linking of this process with "nature" also appropriates the basis of sustainable and biological agriculture systems and their association with natural systems.

In contrast, Ecotrust, an environmental nongovernment organization, approached this issue using tomatoes as a metaphor for the food system. *Start a Revolution With a Tomato* (Ecotrust, 2003) was produced as a tabloid and distributed regionally as a newspaper insert (see Figure 1.1). The sidebar on the cover reads, in part:

> The food industry is consolidating at an alarming rate. The top companies producing meats, grains, and other staples now enjoy

Figure 1.1. A tale of two tomatoes.

Note. Reprinted by permission of Ecotrust.

> virtual control over markets for their products. It's gotten to the point where much of our nourishment depends on a handful of giants. And they're shipping foods an average of 1,500 miles to reach your plate, a practice that strains anyone's notion of "fresh." But a quiet revolution is in the air, and we eaters hold the power for change. Retailers are posting more information about where our food is raised. And new ways of buying direct—like farmers' markets—are providing us with tastier and more diverse choices.

The tabloid uses cartoon-character tomatoes named "Traveling Tom" and "Local Lucy" to represent the differences in food quality and impacts on the environment and local economies between tomatoes produced in the global agro-food system and tomatoes grown in a local agro-food system. Traveling Tom is a tomato bred to be picked green, transported an average of 1,500 miles, and gassed to give it a red color. Local Lucy is a locally grown tomato that tastes great and benefits a local farmer and, therefore, a local economy.

The tension between these examples highlights issues of power: how the environment will be exploited and who will control and benefit from its resources and proceeds—or, more specifically, where farms will be, who will be the farmers, where people will buy food, and what sort of food people will eat. This situation lends itself to analysis using an *ecological* perspective used by anthropologists and other social scientists. Before moving more deeply into the issue of global and local food and agriculture systems, an overview of the research and theoretical framework for this book will be informative.

AN APPLIED AND ECOLOGICAL APPROACH TO ENHANCING FARMERS' MARKETS

In his Malinowski Award Lecture to the Society for Applied Anthropology, Walter Goldschmidt (2001) reflected on Edward Tylor's book, *Primitive Culture* (published in 1889), noting, "the sentence that closes the first textbook in anthropology reads as follows: 'Anthropology is essentially a reformer's science'" (p. 423). Apropos of the topic of this book, more than 100 years after Tylor's comment, Lang (1999) declared that there is a need for reformers—including academics, consumer groups, and environmentalists—to build an integrated understanding of food as a system, and to refine and help organize alliances. This process of reform relies in part on applying methods and theory from anthropology and other social sciences to improve conditions for farmers and communities and to influence policy.

Applying anthropology is a problem- and outcome-oriented approach to research. Kottak (1998) observed, "I see applied anthropology…as the use of…data, perspectives, theory, and methods to identify, assess, and solve social problems" (p. 754). Although applied research should be guided by, and often contributes to, progress in theoretical fields, it must be tempered with a clear view of conducting projects with outcomes that improve conditions for people. This requires examination of and reporting on issues that are of crucial value to, and even generated by, a group in need or distress. Increasingly, applied research is conducted by partnerships of social scientists and members of communities, and highlights the need for collaboration and co-learning between researchers and participants (Trotter & Schensul, 1998).

Applied anthropology should not only lead to solutions for groups of people at both the policy and localized levels, it should also be presented in a manner that can be understood by professionals and practitioners representing many disciplines and issues, as well as by people confronted with problems addressed by the research. The need for unambiguous communication of research findings cannot be overly stressed. Allison Brown (2002), a horticulturist with research interests in farmers' markets, published a review of 60 years of research literature related to farmers' markets. She pointed out that while farmers and farmers' market managers are eager for answers to their problems, there is a fundamental deficiency in some social science research on farmers' markets:

> Unfortunately, some social science studies are presented in ways not immediately useful to farmers or market masters, who are

> more likely to be looking for effective market management tips than semiotics. Highly jargon-laden social science papers may baffle growers seeking hints on improving their market. (p. 173)

Certainly one issue encountered by Brown is the tendency by some anthropologists to use terminology borrowed from literary criticism (Kottak, 1998). This trend distances anthropology—a discipline about people—from people who, if not for the misdirected communication, would be interested in what we say. The effort to develop a local food system and all its associated benefits is a multidisciplinary endeavor. Our research and its implications should be accessible to interested practitioners of other disciplines and to food system advocates. We should be brokers of communication among groups, not part of the problem.

Applied anthropologists have an affinity for the use of theoretical approaches and perspectives associated with ecological anthropology. Ecological anthropology (subsumed under environmental anthropology) is a very big tent that encompasses human ecology, cultural ecology, and other "ecological" approaches practiced by anthropologists, sociologists, and geographers. Among these ecological approaches is political ecology, which is actively utilized by applied anthropologists. A commonly used definition for political ecology originated with Blaikie and Brookfield (1987): "...'political ecology' combines the concerns of ecology and a broadly defined political economy. Together this encompasses the constantly shifting dialectic between society and land-based resources, and also within classes and groups within society itself" (p. 17).

For anthropologists, political ecology connects components of cultural ecology and political economy and offers the ability to take wide views of interactions within a system (Greenberg & Park, 1994). It embraces multidimensional (local, regional, national, and global), multiscale (vertical and horizontal), and historic perspectives. This approach emphasizes the connections of a locale to a regional or global level (Brosius, 1999; Stonich, 1995).

Some authors have remarked on the strong ties between political ecology and applied anthropology. As Paulson and associates (Paulson,

Gezon, & Watts, 2003) noted, "from the beginning...political ecology was analytical, normative, and applied..." (p. 206). They go on to describe elements shared by the approach and applied anthropology:

> The means political ecologists have employed to collect, analyze and apply data overlap in vital ways with those of applied anthropology in general. Some of the shared elements include: concern with environmental decision making and conflict resolution; attention to and mutual collaboration with various kinds of social groups and social movements; interest in the distribution of benefits, costs and risks on various scales; and concern with development models and discourses, together with their environmental and social consequences. (p. 212)

The development of ecological approaches in anthropology, including political ecology, has been dealt with in depth in several excellent reviews. For outstanding summaries of ecological approaches in general, including political ecology, see Kottack (1999), Moran (1990, 2000), and Scoones (1999). For reviews specific to political ecology, Paulson et al. (2003) offer a short but excellent overview. In a recent edited collection, Paulson and Gezon (2005) offer multiple perspectives and applications of political ecology. Robbins (2004) presented an excellent, in-depth review of political ecology from a geographical perspective. Here is a brief summary.

Ecological approaches as practiced by social scientists today have their roots directly placed in cultural ecology. Although early cultural ecology had considerable success with small, rural populations, the application of simple ecological models to human societies became problematic (Greenberg & Park, 1994). Newer ecological approaches are an outgrowth of research related to environmental degradation, especially in developing countries. These studies conflicted with earlier ecological approaches and their emphasis on stability and equilibrium. An expansion of scale was necessary in order to investigate what was occurring. According to Kottack (1999), "in the new ecological anthropology, everything is on a larger scale. The focus is no longer mainly on the local ecosystem. The 'outsiders' who impinge on the local and regional ecosystems become key players in the analysis..." (pp. 25–26). In linking local to global, several researchers have identified the multiple levels at play. For instance, according to Jansen (1998), "to understand the process at stake one has to locate the decisions of the 'land manager' within a chain of relations: land manager—other groups in society—state—world economy" (p. 213). Or, as Stonich and DeWalt (1989) pointed out, analysis should include interactions between individual, household, community, municipal, regional, national, and international levels. Lastly, Steiner (2002) divided human ecology into habitat (home), community (our surroundings and interactions), ecological region, nation, state, and nation state.

From this perspective, the "ecology" of a locale is interpreted to be more than land or ecosystem but to include the sets of institutions that enhance and regulate the land and ecosystem. As Bennett (1990) observed, "all modern uses of the physical environment are mediated by *institutions* [italics original]. The concept of institution is absolutely crucial to the problem of environment in the contemporary world" (p. 441). Institutions may include regulatory entities; nongovernment organizations; city, state, and federal governments; and so on.

Paul Robbins (2004) referred to what he calls the "hatchet" and the "seed" of political ecology. "Hatchet" refers to political ecology as critique; "seed" refers to political ecology as "equity and sustainability" research. With regard to the "hatchet," anthropologists and other social scientists have excelled at critiquing systems that are degrading the environment and marginalizing segments of human populations. A good example for the United States is Thu and Durrenburger's (1998) edited collection of critical research related to social perspectives on the impacts of industrial agriculture. This work helped to refocus attention on U.S. agriculture as a worthy area of research for anthropologists. Studies related to the "seed" are less common.

In a related effort, through their incorporation of political economy into biological anthropology, Goodman and Leatherman (1998) offered a "bioanthropological political economy" perspective. They outlined five interrelated issues:

> 1. The importance of…*social relations* through which individuals gain access to basic resources and labor. These social relations are key to resource production and distribution (that is, they are relations of power) and thus are key to forming proximate environments…
> 2. The importance of *links between the local and the global* (macro-micro interconnections). Threats and benefits…are invariably linked to regional and international processes, and how they intersect with local conditions to shape the microenvironment of adaptation.
> 3. *History and historical contingency* are critical to understanding the direction of social change…
> 4. Humans are *active agents in constructing their environments.* Part of this constructing entails resistance and revolution, as well as accommodation and adjustment…
> 5. *Ideology and knowledge, of subjects and scientists alike, are key to understanding human action*…it is increasingly clear how power and resources are maintained by control of knowledge. (pp. 19–20)

This book is, by necessity, not a detailed theoretical study of the political ecology of farmers' markets; rather, it is a problem-oriented study using a broadly defined political ecology as an analytical framework. An important outcome of this study is the adaptation of constructs from political ecology to assist farmers' market organizers and those who work with them to "see" their environment, thus enhancing decision making and strategic planning. A pragmatic approach is vital. Missing are detailed analyses of the dynamics of power hierarchies and impacts of localized agro-ecologies. To fit the goals for this study, these fundamentals of political ecology provide a framework for analysis:

1. A multilevel analysis of influences (local to global) from the natural environment (biophysical) and the political environment (social relations, economics and policy that are mediated by power hierarchies), using contemporary and historic perspectives.
2. The actions of participants result in adaptations or adjustments to influences from the natural and political environments. Decision making plays a crucial role in action.
3. With the aforementioned fundamentals in mind, this investigation is focused at the impacts and adaptations at the local level for learning (i.e., curriculum and outreach) or potential replication by others facing similar circumstances.

A broadly defined political ecology contributes to this framework throughout the book. That is, the whole of the ecological approaches used by anthropologists and other social scientists are relied on as a toolkit for use in conjunction with political economic themes. This approach facilitates examining farmers' markets in the context of both their natural and political environments, and reveals adaptations to influences from these environments. Although farmers' markets are derived from a shared concept of an open-air market specializing in locally produced foods, individual farmers' markets often are managed differently from and even appear dissimilar to one another. In this way, they reflect the specific conditions in which they exist. These conditions include the social and economic conditions in a community; the local subculture; regulations from local governments; the impact on food production of local soil quality, rainfall patterns, and seasonal temperature variation; and the ability of local farmers to produce high-quality foods from these physical resources. There are solid links to the environment via Oregon's diverse agro-ecozones and their influence on food production and seasonality, the open-air adaptation of farmers' markets (exposure of market sites), and associated impacts on market seasons.

Farmers' markets also respond to national and global influences. These include such widely disparate issues as the impact of international trade agreements on market opportunities for farmers, the provision of funds by the federal government for low-income seniors to shop at farmers' markets, and even the appearance of mad cow disease in beef or E. coli in spinach—which drives a segment of the population to seek beef that is raised on grass and spinach that is not processed under industrialized conditions.

In addition, the environment as rural landscape is threatened by the politics of corporatization and commoditization of agriculture and food. If there are no viable markets for small farms, there will be no small farms. How, then, will rural areas appear, and what will be done there? Farmers' markets are one route to viability for small farms. In this way, this research is related to preventing the degradation of markets and livelihoods for some farmers and the potential degradation of community linkages with rural hinterlands resulting from the power of some corporations and public policy. Thus, farmers' markets symbolize resistance, an important component of political ecology.

Embracing Robbins' (2004) "seed" category of political ecology, this book uses political ecology to provide farmers' market organizers and those who work with them a view of their goals, the multiple-level challenges and barriers to reaching their goals, and how market organizers successfully adapt to challenges and barriers. This moves political ecology beyond critique, to become a problem-solving tool within localized organizations. It becomes a means for organizations to see the big picture and potentially to make decisions that maintain their long-term viability.

Two Food Streams Redux: "What if We Looked at the World as One Giant Farm Field?"

During 1996, while the memory of the Farm Crisis of the 1980s was still fresh, the USDA appointed National Commission on Small Farms held hearings around the country. The Commission was interested in the issues facing small farmers in the United States and was particularly concerned with a long-established national trend of declining numbers of small farms and the increasing size of a smaller number of large farms. This trend had always been presumed to be related to economies of scale enjoyed by larger farms. The trend toward larger farms emerged in tandem with a trend toward consolidation in agribusiness; as one got bigger, the other got bigger. However, one of the key conclusions of the Commission was that market forces were not responsible for forcing small farms out of business. Instead, the loss of market opportunities resulted from federal government policies that favored large farms (USDA, 1998).In Britain, Lang (1999) echoed the Commission's conclusion, asserting that food systems are the outcome of policy and political choices.

As previously noted, companies like ADM are moving agricultural production and food processing to global levels. According to Friedland (2004),

> Globalization in agricultural commodities means that all three classical components of production—land, labor, and capital—experience this shift, albeit variably. Capital is the most easily globalized; labor has become more globalized...It takes place also in the form of bringing employment to workers as manifested by what is referred to as "contracting out" or "outsourcing." Land as a physical entity is immobile but becomes mobile, in effect, through surrogate locations of production. (p. 5)

Trade has globalized food throughout history. The Columbian Exchange and colonial empires moved seeds, diets, recipes, and products rapidly around the world (Lang, 1999). But what we are experiencing now is different: the system is different, the food is different, and the players are different.

The system is different. According to Friedland (2004),

> Although some foods such as sugar have been *somewhat* globalized for about 400 years and fresh produce—bananas—have been *somewhat* in global circulation for over a century, most foods were, up to World War II, in restricted circulation: agricultural

> products were produced and distributed locally, regionally, or nationally. After World War II, particularly after the 1960s, production-consumption chains began to expand. (p. 5)

The food is different. The priority for the food grown in the global systems is based on the ability to be shipped great distances. "The dominant tendency is toward distance and durability...More rapidly and deeply than before, transnational agri-food capital disconnects production from consumption and relinks them through buying and selling" (Friedmann, 1994, p. 268).

The players in the global food system are different. The companies are vast in size and are not tied to a single national allegiance. Frequently they are wealthier than the countries in which they do business. These corporations are qualitatively and quantitatively different from anything that came before them. The size of these firms has long since passed the limit where economies of scale accumulate. They are huge and have many sourcing options (Heffernan & Constance, 1994).

The scale and power of these companies in the United States alone is staggering. The operation of these companies has a tremendous effect on agriculture here. Hendrickson and Heffernan (2005) periodically calculate what they referred to as the "CR4" for a variety of sectors of agribusiness. The CR4 is the concentration ratio (relative to 100%) of the top four firms in a specific food industry. For instance, 83.5% of the beef-packing industry is controlled by Tyson Foods, Cargill, Swift & Company, and National Beef Packing Company. This is up from 72% in 1990. Nearly two thirds (64%) of the pork packing industry is controlled by Smithfield Foods, Tyson Foods, Swift & Company, and Hormel Foods. This is up from 37% in 1987. Nearly half (46%) of food retailing is controlled by five companies: Wal-Mart Stores, Kroger Company, Albertsons, Inc., Safeway, Inc., and Ahold USA, Inc. This is up from 24% in 1997. A 2002 edition of the report showed that Cargill, Cenex, ADM, and General Mills controlled 60% of terminal grain-handling facilities—which means a large portion of the global market for grains is influenced by four companies (Hendrickson & Heffernan, 2002).

Consolidation at this level means significantly fewer markets for farmers and significantly less competition for the prices farmers receive for their products. The business of agriculture at this scale has negative impacts. Studies of industrial agriculture in the United States indicate that an extremely small number of very large farms and agribusiness firms tend to win, while communities, small farmers, and the environment lose (Bonanno 1994; Goldschmidt 1978; Thu & Durrenberger 1998).

There are also international inequities resulting from the movement of commodity production for Western countries to developing countries. These countries may experience a loss of agricultural self-sufficiency, undermining their food security (Beardsworth & Keil, 1997). To make matters worse, developing countries may, in turn, experience delocalization of their own diet, whereby a significant portion of their traditional diet is substituted by imports from other countries (Pelto & Vargus, 1992). These countries will pay other costs as well. Describing the consequence of a highly industrialized agricultural system, Goldschmidt (1998) pointed out that "it is efficiency driven and unsentimentally profit oriented. It takes no account of hidden costs that ultimately may be borne by the community" (p. 185).

Some visions of the future of farming in the United States paint a bleak picture. A report by the Council for Agricultural Science and Technology (CAST) thoughtfully analyzed the "vertical coordination" of U.S. agriculture (Tweeten & Flora, 2001). It predicts a highly industrialized agriculture integrated vertically and detached from rural communities:

> Tomorrow's tightly coordinated agriculture may be characterized more by a hub, spoke, and wedge configuration. For example, a large livestock-processing plant at the hub will be in close proximity to livestock-feeding operations supplied feed by mills drawing grain and oilseed through transportation and communication spokes delineating crop production 'wedges' covering large areas. The nation's landscape will include few clusters. Farms in the periphery (wedges) will require less and less labor and other local inputs per unit of output and thus will provide diminishing social and economic support for their own local rural communities. (p. 14)

Reacting to these types of conditions, Conner (2004) said that "lowering standards in pursuit of 'competitiveness' may be a good way to sell products in the global economy, but it is less effective in sustaining a place where people would want to live" (p. 27).

In contrast, recent surveys of consumer attitudes show a greater interest in food grown closer to home. An American Farmland Trust (2001) survey revealed that 81% of American voters wanted their food to come from within the United States. In addition, a majority (52%) wanted their food to come from their own state. The researchers identified this preference as a "strong undercurrent in the American marketplace" (p. 10). Another national survey probed consumer attitudes toward globalization (Wimberley et al., 2003, p. 1). The researchers asked, "If the U.S. could buy all its food from other countries cheaper than it can be produced and sold here, should we?" Seventy-four percent said that the United States should not. Seventy-one percent said they would be willing to pay more for food that is grown locally. This type of consumer preference creates opportunities for small farmers.

Two Food Streams Redux: "Start a Revolution With a Tomato"

With the loss of competitive market options, small farmers have difficulty participating in conventional wholesale markets. Prices per unit are very low, requiring production of large volumes to be profitable. Small farmers tend not to have the land area and/or other resources to produce these necessary volumes, and even farms that have the ability to produce large volumes still find profitability elusive.

Products in a wholesale channel pass through many hands and sometimes over great distances. There is no direct contact with or feedback from the end user. Farm gate prices in this system tend to be quite low, but costs are added at each stage of this process as middlepersons add profit made necessary by their role in the system. Unable to utilize wholesale markets and remain viable, many small farmers market their products directly to consumers and food-oriented businesses within their local area. Marketing in this manner can be profitable for small farmers and builds a local food economy that circulates dollars between food-oriented businesses, consumers, and farmers. This localized approach to food production, distribution, and consumption is often expressed as *community*, *local*, or *regional food systems* (hereafter referred to as local food systems). Local food systems are defined as a "collaborative effort to build more locally based, self-reliant food economies—one in which sustainable food production, processing, distribution, and consumption is integrated to enhance the economic, environmental, and social health of a particular place" (Feenstra, 2002, p. 100).

Advocates of and participants in local food systems are often resisting the conventional agro-food system and the aforementioned domination by transnational corporations, tendencies toward industrial forms of food production, and globalizing strategies that exploit the developing world. Indications of consumer resistance include the phenomenal growth in components of local food systems, including farmers' markets. For instance, Wright and Middendorf (2008) included farmers' markets among a "new production/consumption paradigm" and part of "...the aggregation of efforts to reshape the convential agrifood system" (p. 3).

Some studies also indicate consumers are recognizing connections between their food purchases and local farms and the local economy. A random survey of households in two Oregon communities asked why consumers buy local food. Forty-six percent indicated that keeping farmers in the area was "very important" and 44% indicated that supporting the local economy was "very important" (Stephenson & Lev, 2004). Another study of farmers' market customers asked if an item grown elsewhere sold for $1.00 at a grocery store, what would they be willing to pay for a similar item grown locally and sold at a farmers' market? Eighty percent of farmers' market customer indicated they would pay more than $1.00 (ranging from $1.10 to $2.00). The study indicates that customers distinguish between local and nonlocal products and are willing to pay more for local products (Lev & Stephenson, 1998).

Using a bioregional approach, a local food system is equivalent to a *foodshed* (Kloppenburg, Hendrickson, & Stevenson, 1996). Just as a watershed provides water, it is a region that provides or can provide food for a region. A related term for expressing the concept of a local food system or local food economy is the term "civic agriculture" (Lyson, 2004). Civic agriculture describes a system made up of economic and personal relationships within a community. The concept emphasizes community economic development, but it is balanced against the social and environmental objectives of a community.

> The term "civic agriculture" references the emergence and growth of community-based agriculture and food production activities that not only meet consumer demands for fresh, safe, and locally produced foods but create jobs, encourage entrepreneurship, and strengthen community identity. Civic agriculture brings together production and consumption activities within communities and offers consumers real alternatives to the commodities produced, processed, and marketed by large agribusiness firms. (Lyson, p. 2)

Among several important characteristics of civic agriculture, Lyson (2007) observed that farm production is for local and regional markets rather than international markets; food production is integrated into communities instead of being geared toward producing commodities to meet the needs of agribusiness corporations; farmers compete on product quality rather than lowest cost; production practices tend to be labor and land intensive and therefore, small enterprises; and civic agriculture producers generally develop direct market links with their customers.

Local food systems comprise many direct linkages between farmers, businesses, and consumers. These points of contact are often innovative and blend a marketing channel with a relationship between the farmer and the consumer, restaurateur, or other local food business owner. Farmers' markets are one such point of contact.

For our purposes, the movement of food from farms to the consumer's table—"from farm to fork"—defines a food system. It "consists of all the steps involved in putting food on the table (e.g., growing, harvesting, processing, packaging, transporting, marketing, consuming, and disposing)"

(Butler & Maronek, 2002, p. 53). In a food system, food is produced in specific locations and follows pathways within or through food system "layers" to locations, whether nearby or far away. Figure 1.2 illustrates in a very simple manner the layers of food systems. Beginning with the local food system, each layer is larger in scale and complexity. Missing within this illustration is an intermediate-scale food system at a regional level that sometimes is used conceptually with "local." Ideally, there would be multiple local food systems contained within a regional or national food system. Generally agreed-upon terms and definitions are a problem with this concept. For instance, the definition of "local" may mean "within a 50-mile radius" for some, while it may mean "within the state" for others.

Food within a local food system is produced and consumed inside its boundary. National and global food systems produce and source throughout large regions or hemispheres and move food through complex distribution channels to local levels. From the perspective of my home in Corvallis, Oregon, bananas produced in Ecuador are part of the global food system; oranges grown in Florida are part of a national food system; peaches grown in the Yakima Valley of Washington State are part

FIGURE 1.2. Layers of food systems.

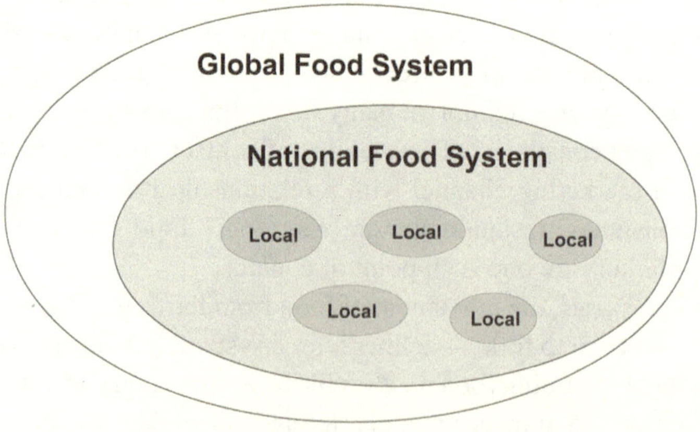

of a regional food system; and vegetables grown at Gathering Together Farm, five miles outside of Corvallis, are part of a local food system.

There are potential benefits to the environment and substantial revenue available for farmers with a more localized food system. An Iowa study examined the distance traveled by 28 fresh fruits, vegetables, and meats via local, regional, and conventional (national and international) distribution systems. The study reported that food grown and consumed in Iowa traveled an average of 45 miles from farm to point of sale. This compares to 1,546 miles for the same items through the conventional domestic system. According to estimates in the report, in addition to significant savings in fossil fuels and greenhouse gas emissions, if Iowa farmers could capture just 10% of these sales from out-of-state sources, they would gain $54.3 million in additional income (Pirog, Van Pelt, Enshaven, & Cook, 2001).

Table 1.1 summarizes the tensions between more globalized versus more localized food systems. The characteristics of the emerging global food system include food traveling long distances, going through multiple custodies, low prices being offered to farmers, and large farms and industrial farming methods being favored. In this system, communities

Table 1.1. Tensions between local and global food systems.

Local	Global
Small farms	Large farms
Diverse farming methods	Industrial farming methods
Enhances rural communities	Degrades rural communities
Low environmental costs	High environmental costs
Food	Commodity
Freshness and flavor principal goals	Shipping and storage principal goals
Farmers	Functionaries
Neighbors	Strangers
Cooperative	Corporate
Unconventional	Standardized
Seasonal	No season
Decentralized	Centralized
Self-reliance	Export-oriented

Sources. Beardsworth and Keil (1997); Ikerd (1998); and Lang (1999).

export their food dollars, and consumers and communities have no linkage to food and farmers and associated rural areas. In contrast, the characteristics of a local food system include food traveling "short" distances with few custodies, more equitable income for farmers, opportunities for small farms, and the encouragement of diverse farming methods and economic and community development via a web of relationships between consumers, farmers, and food businesses.

It is important to point out that completely replacing the current food system with one which is entirely local is neither necessarily realistic nor entirely desirable. From a dietary perspective, some important foods we obtain through our national food system can be grown only in particular climates (e.g., citrus) or are best adapted spatially to particular regions (e.g., cereals and some types of livestock production). There will continue to be a need to link food grown in the vast farming and ranching areas of the American Midwest and West with people in population centers. International markets also provide important and exotic contributions to our diet and market options for farmers that are larger than the scale appropriate for using direct sales. The key issue is shifting the dominance of the agro-food system from one that is corporate driven to one that is community driven.

Farmers and consumers participate in a local food system when they sell into or buy from farm-direct marketing channels. Farm-direct marketing circumvents the conventional wholesale system by eliminating the middlepersons that populate wholesale marketing channels:

> [Farm-] direct marketing allows the farmer to capture a higher proportion of the food dollar. Instead of competing directly with large agribusiness firms, local growers are able to carve out a local market niche and thereby increase their financial viability. [Farm-direct marketing] allows growers to redefine the entire marketing mix of product, price, placement (channel of distribution), and promotion. (Kambara & Shelley, 2002, p. 5)

In addition to the attractive financial aspects of farm-direct marketing channels, the farmers and customers have the opportunity to meaningfully interact. According to Hinrichs (2000), "Direct agricultural markets promise human connection at the place where production and consumption of food converge, an experience not available either to consumers shopping at 'super stores' or 'hypermarkets' or the farmers selling through conventional wholesale commodity markets" (p. 295).

Farm-direct marketing is not necessarily an easy method of marketing for farmers but it is one of the few ways for small farms to be viable. Direct marketing is very prominent in several studies associated with small farmers. A North Carolina study found that small organic producers marketed exclusively through multiple direct channels (each farmer used a minimum of four outlets) (Andreatta, 2000). This relationship between small farms and farm-direct marketing is further illustrated by findings from a California farm-direct marketing study (Kambara & Shelley, 2002). The researchers found the following:

- Smaller farms were more dependent on farm-direct marketing.
- Seventy-eight percent of the farmers stated that farm-direct marketing was important or very important to their operation.
- Sixty-three percent of the farmers reported higher per-unit profit margins from farm-direct marketing sales relative to conventional marketing channels.
- Farmers' markets were the predominant channel used, with 80% of the participants employing this method and 54% employing it exclusively.
- Farm-direct marketing serves as a vehicle for new farm entry and micro-agriculture viability. It was particularly important as a means for small farms to begin operation.

The Kambara and Shelley study illustrates the importance of farm-direct marketing for small farmers and, with relevance to this book, the prominent use of farmers' markets as a marketing channel. In fact, the researchers observed "a large percentage of small direct marketers believed that they really had no choice but to market directly to consumers if they wanted their farm to survive" (Kambara & Shelley, 2002, p. 18).

Examples of farm-direct marketing currently used are U-pick farms, on-farm markets or roadside stands, farmers' markets, community-supported agriculture (CSA), farm-to-institution connections (schools, hospitals, etc.), and farm-to-retail connections (restaurants, retail food markets). U-pick (or pick-your-own) farms and roadside stands have been a mainstay for sales of locally grown produces for decades. Farmers' markets have also long been a part of the American landscape, and because of their popularity, they currently occupy a critical point in newly emerging local food systems. Farmers' markets are recognized by many as "keystones" in the development of local food systems. According to Gillespie et al. (2007), "farmers' markets join together and support seemingly separate social and economic building blocks: the local resource bases and skills of producers, the needs and preferences of local households, and the development goals of communities" (p. 66). As illustrated in Figure 1.3, farmers' markets are places where

FIGURE 1.3. Farmer interacting with farmers' market customers.

Note. CSA is a more recent innovation in linking locally produced food with people.

"market transactions and non-market interactions are joined" (Gillespie et al., 2007, p. 66). Community supported agriculture is a more recent innovation linking locally produced food with people.

> Community-supported agriculture consists of a group of individuals or families who commit resources (money and/or labor) to a farmer and become, in essence, shareholders of the farm. In return for their investment, the shareholders receive part of what the farm produces that season. CSA shareholders provide farmers with the money they need to finance their operations before the growing season begins. In this way, the shareholders incur along with the farmers both the risks and the benefits of food production. (Lyson, 2004, p. 88)

In addition to an advance payment for food, some CSA arrangements often require members to work in some manner to support the farm, an activity meant to build community. CSA shareholders receive a weekly box containing a variety of vegetables, fruits, and sometimes other items that change with the season. Also often included is a newsletter that lists the contents of the box, recipes for using items that are sometimes unfamiliar to the shareholders, and information or news about the farm. Figure 1.4 shows a typical CSA box during the summer season.

FIGURE 1.4. Contents of a weekly CSA box.

It includes potatoes, summer squash, salad tomatoes, carrots, onions, garlic, green pepper, parsley, green beans, leeks, cucumbers, lettuce and blueberries, as well as a newsletter for that week.

Farm-to-institution sales include situations where local public or private programs that require food service for large numbers of people arrange to make some of their purchases from local farmers. The most rapidly developing schemes are with schools and colleges. Farm-to-retail sales offer opportunities for farmers to sell to independent grocers or alternative grocery chains. Both of these types of businesses cater to customers who seek out or appreciate local foods. Many independent, upscale restaurants seek out locally grown foods. A large number of these restaurants embrace the concept of *terroir*, meaning that the region in which food is grown bestows uniqueness on the meal. Terroir is a part of the philosophy of the Chefs Collaborative, a national organization of chefs that support local farmers, ranchers, and independent fishing operators. This attitude is highlighted on the menu of a locally owned restaurant in Corvallis, Oregon:

> As enthusiastic citizens, we care about what goes into our food, and about the impact our buying decisions have on the local community. We continue to build relationships with local organic farmers to bring you fresh, seasonal fruits and vegetables. Our seafood purveyor works closely with Oregon's independent fishing fleet. We prepare only prime and choice natural Oregon beef without added animal protein, hormones or antibiotics. (Big River, 2006)

This demonstrates the potentially robust linkages within a local food system and Lyson's (2004) concept of civic agriculture. The aforementioned restaurant supports local farmers and fishers through its purchases. Customers are drawn to this restaurant because it purchases its food in this way (or perhaps because they like the taste of the food). Restaurants like these are making a commitment to their vision of how food should be produced and served, and the stream of dollars flow accordingly from customer to restaurant to farmer, enhancing the local economy.

FIGURE 1.5. The economic and social stream in a local food system.

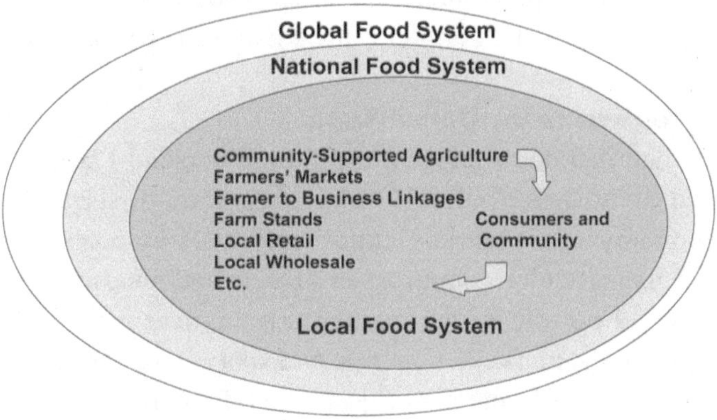

Figure 1.5 illustrates the potential economic and social stream in a local food system. It contains some of the components of a local food system previously discussed. Through the channels in this system, there are dollars circulating and meaningful interactions among the participants. This condition enhances both the local economy and community life. At the same time, this localized system is not entirely independent of national and global food systems.[1]

The Research Literature of Farmers' Markets

The aggregate literature related to farmers' markets is extensive and has many subdivisions. The literature beyond the concerns of this study includes feasibility studies for specific markets (e.g., Anderson, 1995); economic analyses of farmers' markets (e.g., Brooker, Eastwood, & Gray, 1993); how-to guides produced as local Extension Service and Land Grant University publications (e.g., Hodges, 1997); books on the same topic published for national audiences (e.g., Corum, Rosenzweig, & Gibson, 2001); general popular overviews of farmers' markets (e.g., Festing, 1998); and work linking farmers' markets with community food security (e.g., Fisher, 1999). Researchers interested in farmers' markets have also benefited from the periodic appearance of excellent review articles and bibliographies including Lockwood (1972); Smith (1972, 1979, 1980); Pyle (1971); Plattner (1989a); and Brown (2002).

Farmers' Markets in the United States

A superbly detailed description of the East Cleveland Farmers' Market was created by Melvin Smith (1960) as a doctoral dissertation in agricultural economics. It is reminiscent of early work by economic anthropologists in its meticulous counting of objects and measuring of things. Painting a vivid portrait of the market at a moment in time, the study includes data on the market (economic and spatial), its farmers (the market was composed of farmers, not resale merchants), the neighborhood, and the customers. It includes a moderate amount of social data and offers a detailed list of all the products sold and their quantities, the type of dwellings its customers own, the age of customer heads of households and grower and customer opinions of the market, as well as a comparison of shoppers and nonshoppers living within and beyond a two-mile radius of the market. Interviews with vendors and customers helped to generate a list of 11 recommendations for starting a farmers' market. The list is surprisingly relevant today. Here are some examples:

- Is there enough potential volume of fresh fruits and vegetables available?
- Are there three or four growers in each product group interested in developing a farmers' market?
- Are there 15 or more growers available from all product groups—fruit, vegetables, poultry and flowers—to create a healthy market atmosphere?
- Can adequate marketing facilities be provided at reasonable prices?
- Can adequate parking facilities be secured along with the selling or stall facilities?
- Are you familiar with the requirements for [the] physical layout of a farmers' retail market?

- Does the neighborhood have the type of families that would patronize a farmers' retail market?
- Are there enough potential customers to support a farmers' market?
- What market days and hours would be satisfactory for consumers in the neighborhood? (Smith, 1960, pp. 131–132)

Jane Pyle's (1971) "Farmers' Markets in the United States: Functional Anachronisms" offers an excellent compilation of the literature related to development of farmers' markets in the United States leading up to the late 1960s. Pyle's application of the term "functional anachronism" to farmers' markets made both Plattner (1978) and later Brown (2002) recoil, keeping Pyle's work alive for more than 30 years.

Stuart Plattner used the Soulard Market in St. Louis as a research site for several studies (Eckstein & Plattner 1978; Plattner 1982,1984). The Soulard Market was a public market in which both wholesale and retail sales of produce occurred. The vendors were not farmers but merchants who purchased wholesale produce and sold it to retail customers. Based on his experience at the Soulard Market, Plattner reacted to Pyle's (1971) article by arguing that instead of being functional anachronisms, public markets are "functional necessities." He observed that "the public market serves as a 'shock absorber' for the modern produce distribution system by allowing variable grade produce to be sold at reduced prices to consumers instead of being thrown away" (Plattner, 1978, pp. 445–446). He went on to say that "the petty merchants in public markets who substitute labor for capital in the classic style of poor economic actors, provide significant economic benefits to consumers as well as produce producers and wholesalers" (p. 446).

One of Plattner's important contributions was his close examination of the transaction between market merchants and customers and the decisions underlying it. He noted that economic relationships seemed to be as much social as economic (Plattner, 1989b). He outlined two ways to engage in transactions: impersonal (or *atomized*) and personal (or *embedded*).

> Impersonal transactors have no relationship with each other beyond the short term of the exchange; thus in the aggregate they are *atomized*, or not organized into group or social structures. Personalized transactions are between people who have a relationship that endures past the exchange; they are *embedded* in networks of social relations. (Plattner, p. 210)

Plattner maintained that produce merchants "rely on custom and habit, as well as on market information" to make their decisions about what to sell (Plattner, 1982, p. 416). He observed that short-term profit-maximizing behavior was suppressed in order to maintain long-run relationships with consumers.

Anthropologists began turning their attention to urban areas in industrial societies during the late 1960s, and DeWeese's (1974) dissertation on Detroit's Eastern Farmers' Market is a product of that period. With a single market focus, an artifact of anthropologists traditionally limiting their work to a village or other bounded social group, the study traces the adaptation of the Eastern Farmers' Market to changing social conditions and explores the particularistic character of market social relationships. She noted the adaptability of the market and its persistence as it changed functions from a wholesale market to a retail market. From our contemporary perspective, DeWeese was witnessing the rising popularity of farmers' markets. She observed, as Plattner later would, buyers and sellers cultivating personal-exchange relationships entailing reciprocal obligations. Interviews with market customers indicated that their motivations were price, atmosphere, and the freshness and quality of the products—motivations echoed through consumer research at farmers' markets for decades since.

A bookend to DeWeese's study and the decade of the 1970s, Davis' (1978) dissertation, *Small Farmers and Their Market: Relic of the Past or Option for the Future*, focused on a farmers' market in the same decade. Actually, the study is more about farmers and less about the farmers' market; it uses case studies of a number of small farms to highlight the challenges farmers face. The list of challenges sounds analogous to those used here: fluctuating wholesale prices, rising production costs, and changes in food retailing (supermarkets). Davis saw the farmers' market as an adaptive response to these conditions. The work reflects its time, with references to high inflation and rising energy costs, leading Davis to question the long-term adaptability of the U.S. food system. He concluded that farmers were at a crossroads of using farmers' markets as an option for the future. Both DeWeese and Davis produced studies that emphasized the adaptability of farmers' markets. As in-depth ethnographic studies of farmers' markets, they have particular significance at this time of wider-spread interest in the topic.

Appearing in the late 1980s, Sherry's work (sometimes with colleagues) crisscrosses the worlds of anthropology and business marketing. On reconciling the two worlds, Sherry stated, "the use of anthropology to interpret marketplace behavior would assist in applying participating consciousness to a phenomenon that profoundly affects the shape of our lives" (Sherry, 1987, p. 301). Sherry's (and his colleagues') most important work for our purposes was at the "Midville" Farmers' Market. The Midville work included a photo essay of the market and an ethnographic study. The photo essay used a total of 1,377 photos to capture the significance of the opening day of the market (Heisley, McGrath, & Sherry, 1991). This exercise characterized customers by the time of day they attended the market. It explored the relationship between researchers and "informants," and the effect of the presence of researchers (presumably with cameras). The authors observed,

> It was over the long term that we appreciated the cyclicality of the market day and the seasonality of the market itself. By the end of the market we appreciated the relationships between vendors, customers and vendors, and between customers; we found ourselves caught up in the relationships themselves. We ended the experience having grown with and become part of the market itself. (p. 79)

The second portion of the Midville Farmers' Market study includes some sound observations about the market and a disturbing proposal for use of the data. Following a dialectic-tension theme, "Farmers' markets belong to a class of marketplaces experienced by consumers in a very particular way. The structure of such markets unfolds along the dimension of the formal-informal dialectic, and the function along that of an economic-festive dialectic" (McGrath, Sherry, & Heisley, 1993, p. 283).

The authors identified four themes within the market: activism, authenticity, artificiality, and ambience. They concluded that there are "a number of successful lessons our folk merchants can share with traditional retailers" (McGrath et al., 1993, p. 313). Among the many threats to direct marketing channels and the products that are developed within those channels is the appropriation, by more commercialized interests, of what is unique. In their zeal to improve the position of "traditional retailers," Sherry and colleagues seem naive about the potential consequences of their actions for small farmers.

Allison Brown (2001, 2002) has produced well-researched and current reviews of the farmers' market literature. One review (2001) focussed on market inventories and classifications, while the other (2002) reviewed the research related to farmers' markets. The 2002 publication is a well-crafted examination of the farmers' market research literature for the period of 1940–2000. In it, Brown discussed the problems related to data collection on farmers' markets (particularly with the Census of Agriculture), and methodological problems such as differing definitions of direct marketing and reporting conventions. In addition, she summarized the literature on the farmers and consumers of farmers' markets, the economic impacts of farmers' markets, and the social and political impacts of farmers' markets. Discouraged, Brown (2002) referred to the research literature, saying it is "very meager and on critical reading proves somewhat disappointing" (p. 167). Here is Brown's nutshell of 30 years of research findings:

> Farmers use farmers' markets because they feel that they are the best (or often only) market channel available to them. Consumers patronize farmers' markets because they enjoy them and feel that the markets provide high-quality produce at a reasonable price. People with access to farmers' markets enjoy shopping there and think it is socially beneficial to do so. People do not shop at markets

which are inconvenient or of which they have no knowledge. (pp. 168–169)

Applied Research and Farmers' Markets

Applied research will be most useful by concentrating on ways to sustain farmers' markets. A pathway toward this goal is to focus on two complementary aspects of farmers' markets: the external (related to public policy) and internal (related to management) aspects. To improve the sustainability of farmers' markets, public policy must be crafted to enhance the prosperity of farmers' markets. Advocates of local food systems and farmers' markets need data in order to impact public policy; pleas and anecdotes will not change the food system. Economic and social impacts of farmers' markets, preferably quantified, catch the attention of decision makers. At the same time, improving the management of farmers' markets by identifying and resolving their internal issues is another way to improve sustainability.

Demonstrating the use and importance of farmers' markets is valuable for communication with decision makers. For instance, research from two communities in Oregon revealed farmers' markets are patronized, at least on a limited basis, by a significant portion of the population. A random survey of households in the two communities showed that 46% had visited a farmers' market one to nine times during the previous year, and 13% had visited more than 10 times (Stephenson & Lev, 2004).

Documenting the uniqueness of farmers' markets as social and economic units is also important. For instance, another Oregon study indicated that although a significant number of shoppers attended the three markets in the study for the agricultural products (54%, 54%, and 40%), a similarly high number were attracted by agricultural products and market atmosphere equally (45%, 43%, and 53%) (Lev & Stephenson, 1998). A farmers' market's primary reason for existence is fresh food. However, the unique atmosphere is a key aspect of farmers' markets and is something that differentiates farmers' markets from retail food venues. Many consumers choose to shop at farmers' markets because of the unique combination of high-quality products and atmosphere.

On the economic level, advocates and researchers have been interested in the role of farmers' markets for small-farm viability, as business incubators, and as engines of local community economies.

> As a social institution and as social organizations, farmers' markets are important and inadequately studied contributors to local economies. As bridges between the formal and informal sectors of the economy, they enable individual entrepreneurs and their families to contribute to the economic life of their local communities by providing goods and services that may not be readily available through formal mass markets. (Lyson, Gillespie, & Hilchey, 1995, p. 112)

The value of farmers' markets as venues for small farmers has been well documented. For instance, a North Carolina study examined the use of direct-marketing channels by organic farmers. Over 40% of the farmers in the study relied on farmers' markets for more than 75% of their annual income. This indicates the importance of this marketing channel to a significant number of farm households (Andreatta, 2000). An Oregon study analyzed the largely positive impact on seniors and farmers' markets of the federal Senior Farmers' Market Nutrition Program. The results demonstrated a positive economic benefit to farmers selling at small markets in rural areas, despite the fact that the majority of the nearly $1 million in coupons were redeemed in more urbanized areas (Saylor, 2003).

Two studies have examined the business incubator aspect of farmers' markets. A study of California farmers' market vendors revealed the following:

> About 7% of vendors in each market are expanding their businesses each year. Although most vendors simply add a new market, managers reported that farmers' markets provide important contacts for other marketing venues such as restaurant or retail accounts and CSA startups. In addition to new marketing venues, farmers' markets are providing an important environment for developing and testing marketing new products (especially value-added products). (Feenstra & Lewis, 1999, p. 29)

Using the adoption of six practices related to selling at farmers' markets as indicators of innovation and business expansion, another study found that on average, vendors engaged in just 2.07 of the practices—a condition the authors termed "modest." About 24% reported that selling at a farmers' market had helped increase their sales in other outlets (Hinrichs, Gillespie, & Feenstra, 2004). The two studies indicate that farmers' markets nurture some new businesses and offer established businesses opportunities for expansion, but do not do so strongly.

Another economic aspect of farmers' markets is the value of "spillover" sales when farmers' market shoppers also make purchases at nearby businesses. Through the use of a rapid-assessment approach to collecting data at farmers' markets, developed by Lev and Stephenson (1999, 2002), the impact of spillover sales for farmers' markets in several Oregon cities has been documented. These studies indicated that farmers' market customers were drawn into downtown areas because of the market and that they stayed downtown and did additional shopping after they attended the farmers' market (Lev & Stephenson, 1998, 2001; Lev, Stephenson, & Brewer, 2002).

Research directly related to resolving farmers' market management issues or providing guidance for management planning is scarce. One of the few existing studies (Andreatta & Wickliffe, 2002) examined the management challenges at a large, state-operated North Carolina farmers' market. The market depends on seven "anchor" vendors who sell at the market on a daily basis. The anchors provide the market stability because product is available 7 days a week. The market required only 51% of produce sold by a vendor to be grown by the vendor, so the large anchor vendors were engaged in buying large amounts of produce from other venues and reselling it at the market. Small producers could not match this supply. "Because of the market's dependence on the anchors for stability, there is a perception among small farmers that these vendors receive preferential treatment and the market rules are designed to favor anchors, even if this disadvantages small farmers" (Andreatta & Wickliffe, 2002, p. 174). Small farmers suggested that the market was becoming a barrier to their market efforts and interactions with customers.

Their comments on a survey "suggested that policies governing the operation of the market were more important to their satisfaction than the food preferences or spending habits of consumers" (Andreatta & Wickliffe, 2002, p. 174). The market management was sensitive to the need to attract larger producers to ensure stability, but their way of meeting the need "fosters suspicion and dissatisfaction" among small, local farmers. The authors noted, "there may be an inevitable conflict between two possible goals for a farmers' market—to provide consumers with fresh farm produce seven days a week and to provide local small farmers with a marketing outlet" (Andreatta & Wickliffe, 2000, p. 173).

The authors resurrected a useful framework from a Cooperative Extension Service publication from the 1980s. The framework outlined three developmental stages of farmers' markets (Lloyd, Nelson, & Tilley, 1987, p. 1):[2]

- **Stage One:** This is a volatile period which lasts the first couple of years. This stage is unstable and farmers find it too great a risk to sell at the market regularly. Consumers also may be reluctant to support the market if they find few farmers and limited fresh farm products.
- **Stage Two:** A mix of part-time and full-time commercial farmers and home gardeners supply the majority of fruits and vegetables. The presence of the larger producers lowers the chance of market failure.
- **Stage Three:** This is achieved when customers shop at the farmers' market weekly and the market supplies a substantial amount of produce.

Although this is a handy scheme that still has some usefulness, by expanding the focus, Andreatta and Wickliffe (2002) concluded that the market is more than just produce. They noted, that if markets are viewed solely as a way to bring food to consumers, then the framework captures a number of important issues. "But if we want to think of the market as an element in the cultural dynamic that shapes the relationships between farmers and the community for which they produce food, we must recognize that other features of its operation are important" (Andreatta & Wickliffe, 2002, p. 174). The investigation and resolution of farmers' market management issues is, of course, the purpose of this book.

ENDNOTES

1. For critiques of aspects of local food systems, see Andreatta (2000); Born and Purcell (2006); Cone and Myhre (2000); DeLind (1993, 2003); Goland (2002); and Hinrichs (2000).
2. Other land-grant university publications utilized the same concept of stages of development at about the same time (e.g., Thiemann & Burt, 1983).

August Vignette

(*continued from p. 2*)

8:45 a.m.
Mark, of "Red Top Melons," says this is the best crop of melons yet; he's trying a new variety. He asks, "Do you see what they are selling at Safeway for $3?" I help him move his canopy. "This part is bent so I have to mess with it to get it to work. But it's hard to find one with a red top. Gotta have a red top," he says. He forgot his hand truck, so I lift several totes and place them where he tells me. Mark says he can't hump the boxes of melons out of the field. He overfilled them at the start of harvest. He asks another farmer if Miguel and Hector are available. No, they were working for Marlon but were available on weekends. The market is about to officially open. It will run until 1:00 p.m.

9:00 a.m.

A senior couple walks holding hands; a copy of the *Oregonian* folded under one arm shows a portion of a headline "…for Kerry." Dogs. A yelp, then tails wag.

"Spring Mountain" has a selection of leafy salad greens, tomatoes, bell peppers (red and yellow), Anaheim peppers, onions, potatoes, eggplant, celery, fennel, basil, and garlic.

"Donson Farm" has broccoli, cauliflower, tomatoes, several colors of bell peppers, several varieties of chili peppers, potatoes, salad mix, melons, watermelons, summer squash, green beans, shallots, and late-bearing strawberries.

The Democratic Party sets up a canopy just outside the market to intercept voters to register and to recruit volunteers.

10:00 a.m.

"Free corn. Have some more!" Passersby smile. This week the Boy Scouts occupy the community booth, with free corn-on-the-cob. A sign next to the booth says:

> Next Week: Audubon Society
> 'Feeding birds in fall and winter'
> Plant pest experts.

Groups walk, hands in pockets, conversing, waving to acquaintances. Women holding babies stop and greet each other.

"How much does a box weigh? And they cost…?"

A woman quickly turns and adjusts pace as she zeroes in on a stack of tomatoes in blue bins. Plastic sacks say, "Thank You."

A woman says, "*Now* who's tan? You look great!"

In one stroll through the length of the market, I count 16 groups of people standing and conversing.

10:08 a.m.

From the back of the vendor stall, flats of berries are stacked by type: Marion, raspberry, blueberry, and empty. Seven of the 21 large flats are empty.

A child plays an Irish folk song on a fiddle just outside the market. People stop to watch.

10:17 a.m.

Seventeen empty flats of peaches. Seventeen full. The line is five deep to buy them. Farmer wears back support—"Schiek" brand—with shorts.

A dog barks. A man on a bike balances a cedar tree in a 5-gallon pot.

A corn vendor stands on a rear tire to reach into the bed of the truck and arrange the pile of corn.

> "Fresh Corn
> White triple sweet
> Last Day
> Picked Daily"

Kids behind a meat vendor stall slowly roll on a beach towel on the pavement. One has crayons; the other has a game. One is the son of a meat vendor; the other the daughter of a produce vendor in the next stall. They make indiscernible noises. The vendor reaches into one of six ice chests and pulls out a frozen package of ground beef.

Old men in bib overalls line up for peaches.

A couple uses Senior Farmers' Market Nutrition Program (SFMNP) coupons for a single head of broccoli, a head of lettuce, and cherry tomatoes at one of the organic vendors. The transaction is slow and a second employee checks out several customers during the same period.

A man stands under the "Great Roots" canopy: "So, you take…and put it in…and they call that…" He gestures slowly with his hands as two people listen.

10:40 a.m.

The sun is warm as it occasionally breaks completely through the thin, high clouds. Rain is predicted for tomorrow: the first in over a month.

A customer removes his black hooded sweatshirt. The checkout line is nine deep at "Great Roots."

A couple samples watermelon, concentrating on the taste. They hold wedges up between thumb and two fingers. A song ends and a few bystanders clap. A man delicately picks up a wedge of watermelon, places almost the entire red part in his mouth, and examines the remaining rind as he chews.

There is a young woman in a tank top exposing a tattoo extending down her shoulder to her elbow. She holds one child and pushes another in a stroller while she examines meat prices.

Another woman picks up a tiny watermelon and says to the man with her, "Look at this!"

A dog barks. The market manager smiles as she talks on a cell phone. "Fresh hot corn!" rises above the noise.

Corn husks stack up beneath a table. An empty pint container is tossed into a pile; plastic bags zip off a roll; an electronic keypad beeps with each entry. A crowd congregates as two small children play with a customer's puppy on a leash.

A customer pulls stalks of dill from a 5-gallon bucket filled with water. The vendor checks after the sale to make sure all the stalks are still in the water.

Sunglasses, Tevas, flip-flops, baggy shorts and T-shirts.

A woman with bags full of produce places her face over a bag of basil and inhales as she walks; her eyes close.

11:20 a.m.

A vendor in an apron walks through the crowd with $5 bill extending straight out of her hand. She makes her way to the bakery vendor. A vendor in the privacy of the back of his stall counts green SFMNP coupons and Women Infants and Children Farmers' Market Program (WICFMP) coupons. The stack of green coupons is at least 2 inches thick.

I count 307 people in the market on a quick walk through. A few thousand adults will shop at the market today.

Mark, with the "Red Top Melons," explains conventionally grown California melons to a customer. He says they're bred to be hard and pretty and they're harvested in sacks and dragged through the field—"bump, bump, bump"—then put into huge trailers and dumped sideways into a packing shed. "A lot of drops."

11:50 a.m.
An old woman with dahlias in one hand eats corn-on-the-cob with the other hand and smiles.

11:57 a.m.
"Morton River" is sold out of eggs. His car won't be able to move until after 1:00 p.m. Some vendor's stalls are visibly depleted. Even the larger vendors are consolidating their produce into fewer baskets.

"Fine Fruits" is almost out of peaches. The vendor says he held some back to sell at Wednesday's market.

12:06 p.m.
The market is noticeably quieter than earlier. I count 182 people on a walk through.

12:34 p.m.
The music has stopped except for the acoustic blues musician. The Boy Scouts with free corn are gone. Mark has only a few melons left.

The dahlia vendor folds up her tables and consolidates the last items into two bouquets.

12:43 p.m.
The market has been at a lazy pace since noon. Vendors sit and stare into space, the result of being up early with no break. Some vendors visit nearby vendors. Some begin a slow, organized disassembly of their stalls.

A man with an empty market bag walks briskly, scanning the vendor stalls.

12:48 p.m.

The last musician packs up. Several vendors walk toward the restrooms. "Spring Mountain" counts the till while an employee arranges totes and folds tablecloths. The manager walks through the market, smiling.

The dahlia vendor gives away the last two bouquets: one to the acoustic blues musician and the other to a small child. They begin packing up in earnest. Ten customers are still shopping at "Gates Farm."

12:57 p.m.

The oyster vendor canopy comes down. "Spring Mountain" begins moving produce from display baskets into plastic totes.

Sounds of metal poles dropping onto pavement, empty boxes dragged across pavement, clicks, thumps, a truck engine starts.

The first vendor truck drives into the market site at 1:02 p.m. A few customers are still wandering and shopping. "Gates Farm" begins folding table coverings as the last customers check out.

Sandy, the baker, strolls by, pointing to her watch. "One-oh-three!" she shouts to another vendor. Then, with an empty plastic bag and money in hand, she tongs salad mix as "Gates Farm" employees dismantle the stall around her.

1:09 p.m.

A berry vendor quickly loads plastic bags with "Donson Farm" tomatoes and melons; he calls to an employee, "Rick, bring another whole flat."

Chapter 2

The Nature of Oregon's Farmers' Markets

For the past 2 decades, a steadily increasing number of farmers' markets have operated during a typical summer week in communities across Oregon. Hundreds of farmers participate each week—and many more over the course of a market season—in what is, for them, a vital marketing channel. Markets appear on streets and in parking lots, parks, and community centers. They serve thousands of customers and provide a pathway to move a wide array of locally produced food from hinterland to town or city. Farmers' markets are intimately tied to the physical environment by their mobile, open-air nature and by their policy of providing food that is locally grown. The influence of the physical environment and resulting adaptations will be explored here and in subsequent chapters.

MOBILE AND AL FRESCO

There is not a recent tradition of permanent, covered, publicly owned farmers' market facilities in Oregon. Typically, Oregon's farmers' markets are mobile: they use public or private space and they set up, operate, and dismantle within a matter of hours. Virtually all of Oregon's farmers' markets are open-air markets. Although they use the same site each market day, the market itself is transient. They are, as one Oregon farmers' market manager described them, "like a city that exists for a period of time and then it's gone again." Oregon's farmers' markets are well adapted to transience and to open-air settings. Because of this, the site's ability to be transformed quickly into a marketplace is a crucial component of a market.

Geographically, farmers' markets are widely but unevenly distributed across Oregon. Figure 2.1 shows the locations of the 50 markets that participated in this study. More than 60% (31 markets) of the farmers' markets in the state are in the greater Willamette Valley in the northwest section of the state (Portland metropolitan and Willamette Valley regions). The high number of markets in the greater Willamette Valley area is related to the concentration of population and additional environmental factors.

Months and Days of Operation

Oregon farmers' markets operate in a range from 12 months of the year to as few as 2 and a half months (Figure 2.2). The most common months of operation are May through October. Factors that affect a market's selected months of operation are:

- Product availability (influenced by the growing season of the region)
- Dominant weather patterns and their effect on the overall comfort of customers and vendors during the market season, relative to the type of site the market uses

FIGURE 2.1. Location of markets participating in the study.

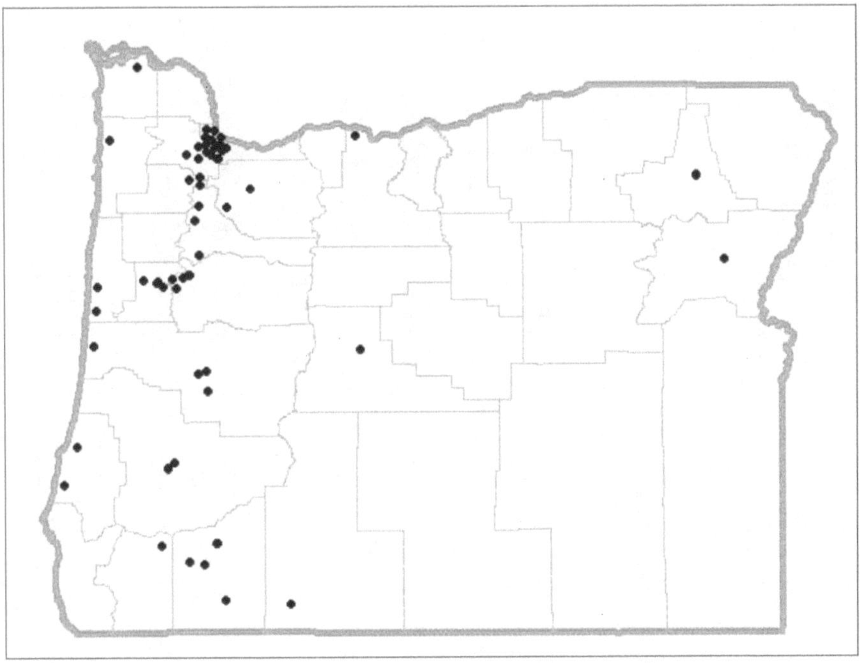

- The ability to target specialized time niches such as afterwork hours

An emerging market niche that affects the months of operation is the increasing popularity of winter or postseason markets. In 2001 there was only one farmers' market operating during the winter. By the 2005 season, there were five markets operating during some portion of the winter months from December until as late as March, and there were three markets that claimed year-round operation.

Other studies of farmers' markets have found similar market seasons. In the mid-Atlantic region of the United States, 84% of farmers' markets opened in May or June. Closing dates were more variable: 50% closed

FIGURE 2.2. Months of operation.

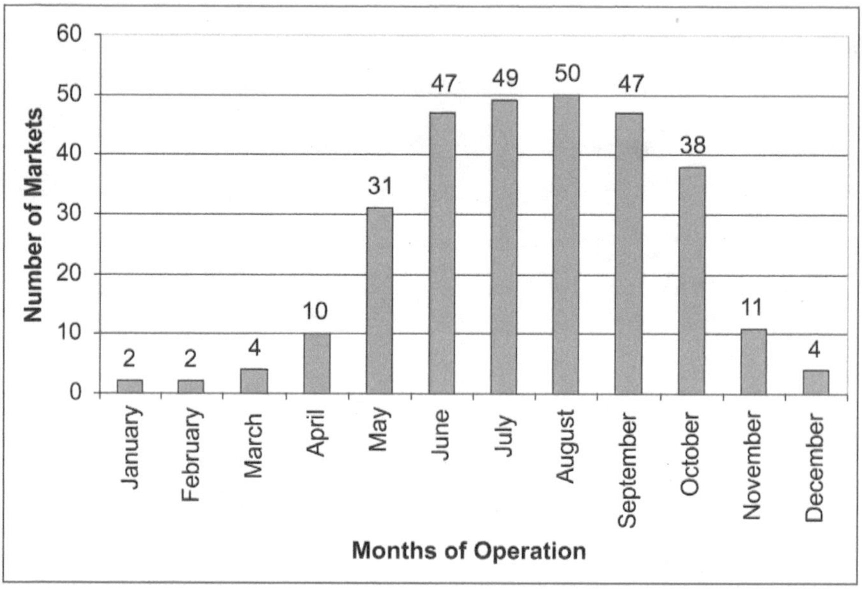

in October, 8% in November, and 18.5% in December. One market in this region was open year-round (Oberholtzer & Grow, 2003). A study from New Jersey indicated that farmers' markets in that area operated from June until the end of October (Govindasamy et al., 1998). These two studies represent northern-tier U.S. states and reflect preferences for operation during the growing season and when it is most comfortable to be outdoors in these regions. A 1994 USDA survey indicated that 14% of U.S. farmers' markets are open all year (Burns & Johnson, 1996). Many of these markets were in areas of California, where a moderate climate and year-long agricultural production can easily support a 12-month farmers' market season.

The day of the week on which a market operates relates to the availability of product, which is dependent on when farmer vendors are available; competition from other markets; and convenience for—and hence availability of—customers. The most popular day for farmers' markets is Saturday, with 62% (31 markets) operating on that day (Figure 2.3). Nearly 40% (19 markets), however, operate on days other than Saturday. These markets often are developed in response to the need for farmers to be able to sell produce more than once a week (because produce is ready to harvest almost every day of the week) and for customers in the community to have more than one day to shop.

For some communities, particularly smaller communities, holding a market on days other than Saturday may be a means to attract a sufficient number of vendors to justify having a market at all. In order to reduce the competition for vendors, and perhaps customers, a market may avoid scheduling on a day when an established nearby market operates. The reverse also may be true. Some markets that are relatively near one another can be timed to be open on the same day to provide convenience

FIGURE 2.3. Days of the week farmers' markets operate.

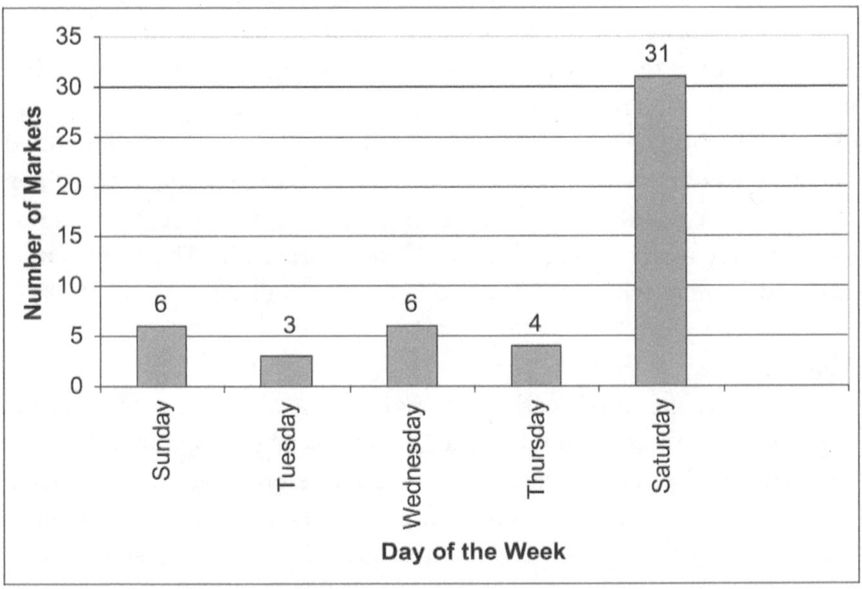

for vendors. These markets benefit from attracting vendors who find the time commitment feasible because there are multiple outlets from which to sell. This situation seems to work best in the Portland metropolitan region because of the very high population base from which the markets draw.

Hours of market operation are established to provide convenient shopping times and in response to environmental issues (such as cold, heat, and wind). One coastal-area farmers' market, for example, sets its closing time to coincide with the time the wind normally becomes problematic. Some urban area farmers' markets are scheduled in the early evening, in the afterwork time slot. Because of the later operating hours, these markets have a short season due to the reduced daylight hours during the earlier and later portions of the market season.

Oregon's farmers' markets operate on days that are largely consistent with findings from a national survey of farmers' markets. It reported that nearly 71% of markets operate on Saturday, with 25% of markets operating midweek and a small number (4%) operating on Sunday (Burns & Johnson, 1996). The survey data for Oregon indicate that 62% operate on Saturday, 26% operate midweek, and 12% operate on Sunday.

Other studies, however, report fewer Saturday farmers' markets. Less than half of the mid-Atlantic region farmers' markets operated on Saturday (44%), with nearly as many operating on Tuesday and Thursday (39%) (Oberholtzer & Grow, 2003). A study from New Jersey identified Saturday as one of the least popular days for farmers' markets. Friday was the most popular day, with seven markets—followed by Wednesday with five markets; Tuesday with four markets; and Thursday, Saturday, and Sunday with three markets each (Govindasamy et al., 1998).

Market Sites

Because most farmers' markets do not own a permanent site, market organizers are faced with finding a publicly or privately owned site that serves the market's needs and can be used consistently. Market organizers select sites based on convenience for customers and vendors, suitability for the size of the market, cost, and availability. Table 2.1 summarizes the primary types of sites used by Oregon's farmers' markets.

TABLE 2.1. Primary types of sites used by Oregon's farmers' markets.

Site Type	Number of Markets	Percent of Markets
Parking lot	23	46
Park	8	16
Street	7	14
Community center	2	4
Fairgrounds	2	4
Other	8	16
Total	50	100

The most common farmers' market sites in Oregon are parking lots, followed by parks and streets (Table 2.1). One of the state's largest markets uses a massive, publicly owned parking lot, whereas a nearby neighborhood market creatively uses both a private parking lot and a city street. Some smaller rural markets use unique sites such as the lawn of a bookstore and private land donated by the market's volunteer manager.

Markets use both publicly owned and privately owned sites and either pay rent or use the site at no cost. Publicly owned sites include municipal parking lots, streets, and parks. Of the 68% of markets (34) that use publicly owned sites, 42% (21) use the site at no cost, and 26% (13) pay rent. Privately owned sites include parking lots or other spaces provided by businesses, churches, and banks. Of the 15 privately owned sites, 2 markets pay rent and 12 use the site at no cost.

Some markets have made progress toward community institutionalization by having sites dedicated for their use. For instance, one market uses a parking lot that provides parking for city employees during weekdays but is a permanent home for the market on Saturdays. There is even a large, eye-catching sign for the parking lot, announcing that it is the "home" of the market. Another market was integrated into the city's riverfront redevelopment plan through provision of a street site that uses ingenious gates to close the street to automobile traffic during market hours.

The use of mobile, open-air, public or private property by Oregon's farmers' markets is consistent with farmers' markets in other geographical regions. A study from Iowa reports that the "vast majority [of farmers' markets] made use of available spaces, such as parking lots, streets or sidewalks," with 70% of markets using public property (Hinrichs, 2001, p. 1).

PRODUCERS AND PRODUCTS

The type of vendors participating in farmers' markets differs among markets: Some markets allow craft vendors; others only allow farmers. Most farmers' markets require farmers to grow most of what they sell and to be from within a region delineated by market rules.

Market Size and Vendor Composition

The size of a farmers' market is defined here as the number of vendors present on a market day during the best-attended months. Although farmers tend to be the dominant vendors at most markets, this is not always the case. Here, market size refers to the market's total size, including all participating vendors (farmer and nonfarm vendors). This is because nonfarm vendors generate revenue for the market and their presence has an impact on market scale and management effort.

Markets during 2002 ranged in size from as few as 5 vendors to as many as 90. By 2006, the upper limit had increased into the low to mid-100s. Although the mean size was about 33 vendors, with nearly half of the markets (48%) have 20 or vendors fewer. Again, this represents the number of vendors present during the best-attended months and includes all participating vendors, not just farmer vendors.

In other areas of the country, as in Oregon, market size is highly variable. Hinrichs (2001) reports a range from 2 to 165 vendors in Iowa markets. Mid-Atlantic states' markets record a large proportion of markets (72%) with fewer than 10 vendors, about 23% with between 11 and 40 vendors, and only about 5% with more than 40 vendors (Oberholtzer & Grow, 2003).

Farmers' market vendors consist of two major types: those who sell agricultural or food products of some type and those who sell craft products. The agricultural or food product vendors (for ease of communication, these are referred to hereafter as "farmers," with the observation that some of these vendors are not farmers in the true sense of the word) can be further broken down into those selling fresh farm products (produce, meats, and cheese); nursery items (bedding plants, flower baskets, and perennials); value-added products (jams, salsa, etc.); and bakery goods (breads, pastries, etc.). Individual vendors often embody two, or even all four, types of product vendor. In addition, many markets include vendors selling ready-to-eat foods as a convenience to customers. These vendors are almost always few in number. In most cases, all the food-oriented vendors must have a "local" connection, such as providing locally grown vegetables, products from a locally owned bakery, or value-added food using locally grown ingredients. Markets have wide and varying policies toward craft vendors. They may require that the products have a gardening or farm use. Market organizers may jury craft products before allowing them to be sold. They may allow any and all crafts, or they may not allow crafts at all. Most markets also offer space to community groups.

Farms Versus Crafts

The presence of craft vendors at farmers' markets is often a strategic consideration for market managers. They may be present because a market has a goal of supporting small nonfarm businesses in the same manner as small farm businesses are supported. In some regions, a crafts-dominated market appeals to the tourist segment of the local economy and provides a venue for farmers that might not exist otherwise. Other markets use craft vendors to add "critical mass" to a small market or to a market during the early or late portion of the season when farmers are scant.

Craft vendors provide a stable revenue stream for markets because they appear each week for the entire season and pay their stall fee. Table 2.2 indicates that most markets allow craft vendors. Only a relatively small number of markets do not allow craft vendors. The remaining majority allow them in varying proportions. About half of the markets that do allow craft vendors have an established ratio of farmers to craft vendors

TABLE 2.2. Oregon farmers' markets and craft vendors.

	Number of Markets:	Percentage of Markets
Allow craft vendors		
Yes	37	74
No	13	26
Total	50	100
Maintain a ratio of farmers to crafts		
Yes	19	38
No	18	36
Do not allow craft vendors	13	26
Total	50	100

that they try to maintain. The ratio of farmers to craft vendors varies widely. Some markets try to maintain 75% farmers, others, 60% farmers. Some markets find it hard to attain their ratios. One market manager commented that the market has a "50:50 percent goal, but actually it is about 75% crafts," indicating a difficulty in recruiting farmers. Other markets do not maintain an established ratio of farmers to craft vendors.

There are differences among markets in their food vendor component compared to their craft vendor component. Table 2.3 indicates that the range of vendors is similar when counting farmers and crafts, as opposed to farmers only (this is skewed to some extent because most of the largest markets are exclusively farmers' markets and do not allow craft vendors). The median, however, is very different. The median size for markets based on total size (all vendors) is 29; however, it drops to 17 when counting farmers only.

Consequently, three patterns for farmers' markets emerge based on the size of their farmer and craft vendor components:

- Farmers' markets (all, or nearly all, farmer vendors)—often referred to as "producer-only" markets
- Farmer and crafts markets (a significant number of craft vendors, but not dominantly so)
- Farmers within a crafts market (craft vendors are dominant)

TABLE 2.3. Market size: Number of vendors.

	Total Market Size: Farmer and Craft Vendors	Market Size: Farmer/Food Vendors Only
Minimum	5	4
Maximum	90	90
Median*	29	17

*Median used here to relieve skewing from variance at extremes of range (*n* = 50).

A higher proportion (74%) of Oregon's farmers' markets allow crafts than do markets nationally. The 2000 national survey of farmers' markets reports that 58% of markets allow crafts (Payne, 2002). The survey of mid-Atlantic states' farmers' markets reports that crafts were the most commonly restricted product, yet 51% still offered them (Oberholtzer & Grow, 2003).

Farmers and Products

It is important to point out that there are more farmers that participate in a farmers' market during a season than are present on any given market day. Many farmers specialize in one or an assortment of crops that are salable only at specific times during the season. For instance, a blueberry grower might be present during only a 4- to 6-week period of a 30-week market season. Most apple growers miss the first several months of the market season and appear perhaps in September or October. So, markets rely on a larger pool of vendors than are present on any single market day. For example, one medium-sized market surveyed draws from a pool of more than 140 farmers, yet has only about 30 to 50 present at each of its markets.

The array of food products sold at farmers' markets is summarized in Figure 2.4. *Fruit* and *Vegetables* includes a rich variety of products and production systems. *Nursery* includes annual and perennial landscape or food plants. *Cut flowers* includes a variety of flowers arranged in bouquets or sold in bulk. *Bakery* includes breads and pastries. *Value-added*

FIGURE 2.4. Products sold at farmers' markets.

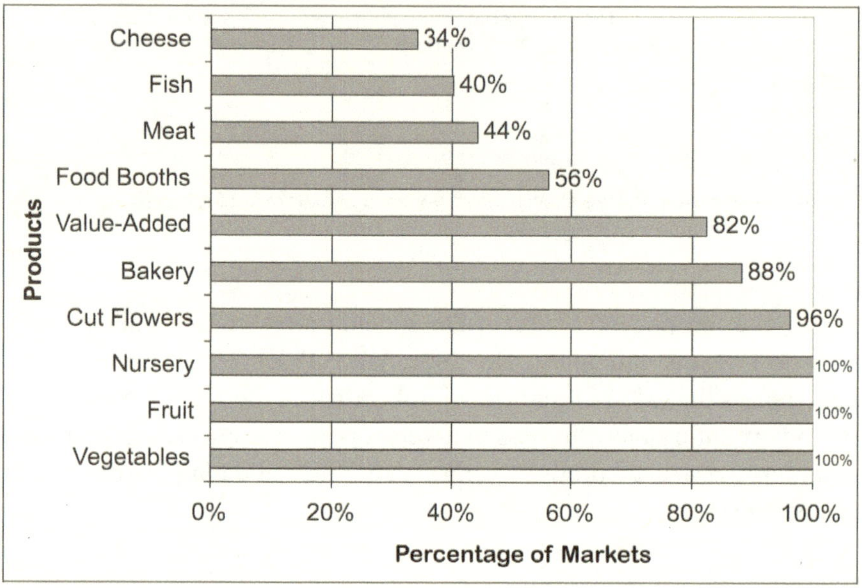

includes jams, salsas, and similar products. *Food Booths* sell prepared foods to eat on-site as a convenience to customers. *Meat*, *Fish*, and *Cheese* include sustainably grown beef, lamb, and poultry; wild fish; and local specialty cheeses.

All markets offer fruit and vegetables, the products for which markets are best known. Recently emerging into many markets and growing rapidly are protein items such as cheese, fish, and meats. These items are sold at fewer markets, but vendors in these categories are in high demand.

Maintaining Localness

Unlike mainstream grocery chains that sell products produced nationally and internationally, farmers' markets connect themselves to a locality by requiring that the food sold in the market be produced by the farmers selling the food and that it originate in an area delineated by market rules or determined by practical driving distances for the farmers. These requirements tie the food to a farm and a region. Consequently, the products in the market reflect what can be grown within the region, and the products change with the time of year and growing conditions. The products vary from crops with short periods to maturity or those that can be grown in cooler springtime conditions (greens and salad mix, for instance), to crops requiring intermediate to long periods of time to maturity and/or requiring high heat units (e.g., corn, peppers, and tree fruits).

Because buying fresh, locally grown produce directly from the farmer is a major attraction of food-oriented farmers' markets, many markets have implemented rules to ensure that participating farmers are, for the most part, growing what they are selling. These rules are intended to maintain the "localness" of the food sold at these markets and to prevent vendors from "high stalling," or purchasing products from other farmers or wholesale outlets and selling them as their own product. For example, in this excerpt from the vendor handbook for one large market, the organizers establish what they consider to be "local" and prohibit products that have been purchased for resale:

> All products must be grown, raised, produced or collected in Oregon or Washington. Products allowed to be sold at the Market are agricultural products such as plants, fruits, vegetables, herbs, flowers, seafood, meat, poultry, honey, processed foods and eggs. All products shall be of good quality and must comply with any applicable regulations pertaining to their production and sale. The Market is for farm fresh locally grown products and is not an outlet for wholesale produce. (Beaverton Farmers' Market, Inc., 2005. p. 16)

Markets that require products to be grown and sold by the farmer indirectly establish the region from which they may draw their food as the distance a farmer is willing to drive. In practice, this may be as much as a several-hour round trip. It is common for farmers from the Willamette Valley to participate at markets in the Central region and Coast region, and the markets in the Portland area draw farmers from the Southern Willamette Valley and the Columbia Gorge.

Table 2.4 shows that nearly 80% of farmers' markets (39) require 75% to 100% of farm produce to be grown and sold by the vendor. A few markets have no rules pertaining to whether the person selling the produce grew it. Three markets do not require the seller to have grown the product, but they do require the produce to be grown within the region.

There is a debate within the farmers' market community concerning the issue of requiring 100% grower-produced versus 75% (or other value) grower-produced product. Some markets that allow less than 100% of products to be grown by the farmer selling them address the resale situation by requiring that the products being resold be identified with the name of the farm where the product was grown. This approach allows some reselling, yet satisfies the "locally grown" goal for the market. It also discourages the practice by some vendors of buying products through wholesale channels and selling them as locally grown.

"Producer only" restrictions are common nationally and are one of the hallmarks of modern farmers' markets. The 1994 USDA national survey of farmers' markets reported that 79% of markets have regulations and restrictions on who can sell and what can be sold at their markets. These markets require farmers to sell only their own crops or restrict farmers

TABLE 2.4. Percentage of farm produce required to be grown and sold by the vendor.

Percentage of Produce Required to be Vendor-Grown	Number of Markets	Percentage of Markets
100	17	34
75	22	44
50	3	6
25	1	2
No rule	4	8
Other:	3	7
Must be from Willamette Valley		
Must be from Oregon		
Must be from 50-mile radius		
Total	50	100

who are selling both their own and other farmers' locally grown crops (Burns & Johnson, 1996). The 2000 USDA national survey of farmers' markets reports that 75% of markets identified themselves as "producer only," which is defined here: "vendors must grow at least some of what they sell" (Payne, 2002). A study from New Jersey reported that 73% of markets do not allow wholesalers or resellers (Govindasamy et al., 1998), and an Iowa study noted that 51% of farmers' markets restrict wholesale and resale (Hinrichs, 2001).

To summarize, Oregon's farmers' markets have both a diversity of adaptation and some common elements. They are unevenly distributed across the state, with most being concentrated in the densely populated and agriculturally productive Willamette Valley. The markets are transient in nature. They are able to set up quickly and transform a stark street or parking lot into an active marketplace, and dismantle just as quickly. A high priority for farmers' markets is a site that is easily converted into a marketplace, is visible to customers, and has sufficient amenities for customers and ease of operation for vendors.

The physical environment is inherent in the nature of farmers' markets. Retail food venues that operate nearly every day have a wide selection of food sold nationally and internationally regardless of the season and are heated/cooled and well lighted, whereas farmers' markets operate in the local natural environment and climatic elements. The open-air form of markets and their requirements that products be locally grown tie markets to the physical environment. The length of the market operating season and times of operation, for example, are influenced by the seasonal amount of daylight and physical comfort for vendors and customers. Climate differences within Oregon influence the length of growing season (such as average frost-free days and accumulation of heat units) and thus affect market operations in terms of crop availability. There is also variability in each growing season based on drought, above-average rainfall, cloud cover, and other climatic influences on plant growth and farming operations.

Rules established by farmers' markets that require the food to be produced within a delineated area tie participating community members to their physical environment by focusing product choices on foods that can be grown in their region. Food sold in this manner embeds community members in the cycle of seasons through the procession of foods available as the growing season advances. Through participation in the seasonal cycle of food production, community members experience, along with farmers, individual years of low and high productivity resulting in scarcity or abundance of specific products at the market.

Chapter 3

Ninety Years of Oregon's Farmers' Markets: An Inventory of Markets and the External Influences on Their Growth and Decline

From 1994 to 2005, a period for which there are reliable inventories, the number of farmers' markets in Oregon increased significantly from 18 to 68. This 12-year trend should not necessarily be viewed as a forecast for the future, because there have been other periods of growth *and* decline during the past 90 years. Examining the historical record reveals how external factors have influenced farmers' markets. A better understanding of these influences may, in the future, enable market organizers to take advantage of positive influences (such as public policy), or assist in shaping them, and to mitigate negative influences on the existence and growth of farmers' markets.

COUNTING FARMERS' MARKETS: AN OREGON INVENTORY

Although it seems like a simple task, counting farmers' markets can be difficult. A number of studies have sought to count farmers' markets in the United States. These studies provided valuable information on the status of markets in their time, but they have all fallen short of providing a complete picture.

The challenge of counting markets hinges on definition and access. How many farmers' markets are identified depends on how one defines a farmers' market. The definition has changed over time and varies from researcher to researcher. Finding markets can also be difficult because their mobility and informality affects their visibility. Finally, some market organizations seemingly do not wish to be found. As a result, many markets were never included in published lists. The information presented here is based on several sources: published resources related to farmers' markets in Oregon, including histories and national inventories; a series of annual directories listing Oregon farmers' markets from 1998 to 2005 published by the Oregon Department of Agriculture (ODA) and the Oregon Farmers' Market Association (OFMA); and data collected as part of a survey of Oregon farmers' market managers and interviews with them.

Growth and Decline

Two prominent researchers have reviewed the status of farmers' markets in the United States at two very different points in time. Jane Pyle (1971) published "Farmers' Markets in the United States: Functional Anachronisms" during a prolonged period of decline in the number of markets. The few markets operating at that time were often considered quaint links to the past. Allison Brown (2001) published "Counting Farmers' Markets" during a period of rapid growth in the number of farmers' markets. This period of growth began gaining momentum shortly after Pyle published her article, and with the exception of a brief decline during the late 1980s, this growth continues to the present.

Both authors documented periods of growth and decline in the numbers of markets over time. Pyle (1971), who extended her study deeply into the 19th century, remarked, "for a variety of reasons, often differing from place to place, the farmers' market waxes and wanes in popularity" (p. 196). Brown (2001), drawing from the perspective of witnessing the late 20th-century upswing in the number and popularity of markets, documented four surges in numbers of farmers' markets nationally:

1. A post-World War I increase in open-air markets replacing municipal markets
2. An increase around 1930 linked to self-help efforts during the Great Depression
3. A late-1970s increase associated with political activism and the Farmer-to-Consumer Direct Marketing Act of 1976
4. The current surge, which began in the 1990s

Even based on the scant literature related to farmers' markets in Oregon, it is apparent that influences operating at national and local levels have affected the number of farmers' markets over time. The rise of the suburbs and supermarkets, political action, a "back-to-nature" movement, the renaissance of cities, and other external factors have all influenced the expansion or decline in the number of markets. There is mostly conjecture and little research regarding what is influencing the current growth in numbers of farmers' markets. This chapter sheds some light on this issue.

The Problem With Definitions

The definition of markets, and therefore the inventory of them, has changed over time. Pyle's (1971) examination of farmers' markets from the 1800s to 1969 acknowledged the difficulty of defining farmers' markets. She pointed out that what is counted as a farmers' market often is not actually a farmers' market. Over the years, terms such as "public markets," "municipal markets," "tailgate markets," and many others have been used to indicate what we might refer to today as "farmers' markets." Brown (2001) discussed the many definitions of farmers' markets used over time in detail and remarks that the farmers' markets of the present are very different from markets in the past: "The nature of farmers' markets has changed over time, and the producer-only retail farmers' markets of today were uncommon until recently. In earlier times farmers' markets mingled producers and resellers, but the function of markets has changed, the definition has become more categorical, and various restrictions on reseller participation have been instituted" (p. 658).

Current U.S. Department of Agriculture (USDA) publications, beginning with Johnson and Bragg (1994), define farmers' markets as "a common facility or area where several farmers/growers gather on a regular, recurring basis to sell a variety of fresh fruit and vegetables and other farm products directly to consumers" (p. 2). Even with the useful and generally accepted definition used by the USDA, there is still a lack of complete agreement on some of the details of the definition.

Published Histories, Inventories, and Reports for Oregon's Farmers' Markets

Inventories for the first half of the 20th century have excluded many open-air markets or less formal markets. Public markets with permanent structures are most commonly represented during this period and often only in prominent cities. The present form of open-air, producer-only markets (i.e., retail sales from farmers, not wholesale operations) is reflected in inventories beginning in the 1970s.

Early 20th Century

There are few written accounts of a historic nature about Oregon's farmers' markets. This being the case, the book *Market Days, An Informal History of the Eugene Producers' Public Market* (Bettis, 1969) is a treasure. Written in 1969, 10 years after the market closed, it offers an abundance of detail on how and why the market formed and follows its growth, facility expansions, and decline. The appendices include lists of all its market masters (market managers), vendors who regularly sold at the market, and other details. Information presented in the book also establishes the existence of a public market in Medford, Oregon, by 1918; a market not documented in early inventories.

Started as a civic project by Lane Pomona Grange of the Patrons of Husbandry, the Eugene Producers' Market began operating on September 4, 1915, in a shed containing 22 stalls "on Park Street on the public square." The market eventually outgrew its first site and moved out of its shed structures in 1929 to a larger concrete building and a new location. Its growth suffered during the Great Depression, boomed again during World War II, and declined for the last time during the late 1950s as supermarkets became the "modern" choice for food purchases. The market closed in 1959 (Bettis, 1969).

A useful book titled *A Market for the City, A History of Portland's Public Markets* (Eigo & Engeman, 2002) follows the growth and demise of Portland's large public marketplaces. The book follows these markets from the 1873 opening of Ankeny's New Market and Theater (also known as the Central Market) and ending with the demise of the Portland Public Market in 1942. Important to this study is the authors' connection of the ebb and flow of Portland's markets to community- and national-level economic and demographic trends over time. For instance, an early public market faded because of a shift in downtown residential districts to areas beyond the reach of the market. Later, commercial decentralization away from the waterfront had an adverse impact on another market. Finally, postwar suburban sprawl led to the closing of the last public market.

Although the book provides an entertaining history of Portland's major public markets, it does not provide a comprehensive record of all public and farmers' markets in the city. It does, however, offer some interesting indications that multiple public and farmers' markets were operating in the city during specific periods of time—depending, of course, on how one defines a "market." For instance, following the closing of the Metropolitan Public Market in 1895, the authors noted, "Chronicles and public records indicate the presence of a number of public and farmer markets operating in Portland over the next two decades, but none of them achieved more than local or fleeting recognition" (Eigo & Engeman, 2002, p. 7). And later, "Throughout the 1930s and 1940s, local producer and farmer markets maintained their own…" (p. 23). But the names and locations of these markets are not offered.

Several national inventories of farmers' markets were carried out at various times during the 20th century. Each of these documents draws only an incomplete picture of the situation because of the immense challenges associated with this type of task. Oregon appears in some of these inventories. For the early period of the twentieth century, the U.S. Bureau of Census conducted a national inventory of "municipal markets" in 1918 (Grogan, Carruthers, Rolle, & Hole, 1919). Limited to cities with populations of more than 30,000 people, the inventory identifies one market in Portland. Eugene and Medford were too small to appear in the inventory and their markets known to be operating during 1918 were not counted.

Mid-Century

In another national inventory, Wann, Cake, Elliot, and Burdette (1948, p. 101) catalogued "farmers' produce markets" operating during 1946. The study reveals two retail markets operating in Oregon. One, the Producers' Public Market in Eugene, is described as having a cooperative management system (rather than private) and records 1916 as the year it began operating, although Bettis (1969) indicates that it began operating in 1915. The second market, the Farmers Cooperative Market in Portland, is also described as a cooperative management system; 1934 is given as the year it began operating.

Eigo and Engeman (2002) noted that the Farmers Cooperative Market was formed as a rival to the huge and well-financed Portland Public Market that also opened in 1934. The Farmers Cooperative Market was the product of growers associated with the much-beloved and municipally owned Carroll Market that was demolished as part of the political battle to make way for the corporate-owned Portland Public Market. In an ebullient history of public markets during the 1930s (including the Portland Public Market), Zimmerman (1937) described the passing of the Carroll Market: "And on the eventful day of December 15, 1933, a new market was opened up and the Carroll Street Market disappeared" (p. 90). Eigo and Engeman documented high levels of debate and sometimes acrimony surrounding the establishment of Portland's major markets. What happened with the Carroll Street Market and the Portland Public Market was one of the most contentious and controversial of these instances.

The Decline During the 1950s and 1960s

There are apparently no inventories of farmers' markets for Oregon following the study by Wann et al. (1946) until the mid-1970s. This 30-year gap is consistent with a lack of inventories for the same period at the national level. Because of the lack of inventories for this period, what is presumed to be the most profound period of decline of farmers' markets is not well documented (Brown, 2001). During the 1950s and 1960s, in Oregon and elsewhere, farmers' market numbers were negatively influenced by a government- and business-led transformation of the food system. As part of the entangled aspects of postwar modernization that included building communities to accommodate the automobile and the convenience of supermarkets, the goal of these transformations was to build an agriculture and food system that was a model of efficiency.

Local accounts record this period. Eigo and Engeman (2002) observed how the stability of farmers' markets during the war years gave way to decline during the 1950s:

> In the wake of the postwar suburban sprawl and the decentralized shopping patterns it signaled, the idea of a central Portland market seemed outdated. More residents fled urban spaces for suburban tracts. Meanwhile, supermarkets and chain stores—both of which promised national brand reliability, easy-term credit, rational organization, and clean and efficient service—claimed customer loyalty and became the model consumer experience. In a world that now valued consensus and homogeneity, the notion of local color and flavor seemed idiosyncratic. (p. 23)

In Eugene, the Producers' Public Market struggled. It finally succumbed, the last straw being the value of its location on the real estate market. Recalling the late 1950s, Bettis (1969) laments the following:

> The market remained. It was a tired, limping memory of its former glory; a great, dim, echoing chamber of yesterdays, inhabited by the last few of those producers who had once made the old building ring with friendship and happy "Good mornings." It was a ghost but it remained. And as long as it remained, so did hope. Maybe it would yet be remodeled. Maybe it would once again be a place of bright displays of produce and the smell of fresh-baked bread and…Maybe…(p. 118)

The market closed on May 4, 1959. "Thrifty Drug Stores of Los Angeles announced that a new drug store and at least one other store would occupy the completely remodeled building by fall. 'Modernization' would begin soon" (Bettis, 1969, p. 118).

From her perspective in the late 1960s, Pyle (1971) noted that the function of markets during this time had changed from their origins, and "arguments for retention of public markets no longer emphasize their function of providing basic necessities to a non-producing population" (p. 167). But, offering some optimism, she inferred that the markets that persisted in 1969 "give evidence of success that may herald still another upswing in fortune of the farmers' market" (Pyle, 1971, p. 197). It was the late 1960s, and there were changes taking place in the thinking of some consumers about food and how to obtain it.

1970s and Early 1980s

According to Brown (2001), a decade of growth of farmers' markets began almost immediately after Pyle's article was published. This was a "result of economic and political activity in support of both farmers and consumers in an era of economic and political turmoil" (p. 657). Although there is not any published material that documents the status of farmers' markets in Oregon during the early 1970s, there was growing interest in farmers' markets nationally. In fact, by the mid-1970s there was sufficient interest from both farmers and consumers in the concept of marketing farm food products directly to customers that Congress began crafting legislation to enhance this type of marketing.

An article from that period in the agriculture trade magazine *American Vegetable Grower* ("Farmers' Markets," 1977) captures some of the excitement: "Spurred on by the 'back to nature' trend, direct marketing lends an advantage to growers who strive to get high quality, ripened-to-perfection produce into the hands of the consumer" (p. 11).

The passage of the Farmer-to-Consumer Direct Marketing Act of 1976 (Public Law 94-463)[1] directed the Cooperative Extension Service to become involved in farm-direct marketing and to include the organization of farmers' markets, among other types of involvement. State departments of agriculture were also directed by the Secretary of Agriculture to enhance this form of marketing. There was funding available to facilitate these efforts. A period of rapid expansion on the national level followed from 1977 to the early 1980s.

During 1976, *American Vegetable Grower* conducted a survey to count roadside markets and, on an incidental basis, farmers' markets in the United States. Responses to the survey were received from 30 states. Oregon was not among them ("AVG Survey," 1977). This survey is important because the next USDA effort to inventory farmers' markets in the United States relied, in part, upon it.

Linstrom (1978), an economist with the USDA, developed an inventory of farmers' markets in the U.S. with data from the aforementioned survey and supplemental information that included other printed information and contacts with personnel at land grant universities and state departments of agriculture. Linstrom noted 12 farmers' markets operating in Oregon at that time. Locations are not recorded. Because Oregon does not appear to have participated in the *American Vegetable Grower* survey, the figure for its number of markets was likely obtained from the other sources accessed by Linstrom.

A master's thesis from Oregon State University (Thiemann, 1983) noted that during 1981, when the research was being conducted, there were "roughly a dozen" farmers' markets in Oregon. It is possible that Thiemann's rough estimate was based on Linstrom's estimate of 12 markets from 1977, although that study is not cited. Thiemann goes on to note that most of the markets were less than 5 years old, and that many communities were planning to establish new markets. This may connect some of the markets to the growth in numbers stimulated by Public Law 94-463.

Decline in the 1980s

Momentum was lost during the late 1980s. The political climate in the nation was not supportive of small farms. The National Commission on Small Farms (USDA, 1998) identified the 1980s as the period of the greatest movement in federal policy toward favoring large farms. Brown (2001) records that "a dip in the number of farmers' markets occurred during the Reagan Administration when funding for community development projects was curtailed" (p. 666). In Oregon, from what little can be reconstructed, the growth of farmers' markets seems to have stagnated or declined slightly during this period.

Inventories From Oregon Farmers' Markets Directories (1998–2005)

In 1998 the Oregon Department of Agriculture began publishing directories listing Oregon's farmers' markets. The directories list market locations; months, days, and hours of operation; and a contact person for further information—often the market manager. Because of publishing deadlines and poor communication within the farmers' market community, in the early years these brochures often did not contain all the markets that were operating during the year of publication. It appears that some markets were not aware of the brochures and were not listed for a year or more after they opened. By 2000, a revitalized Oregon Farmers' Market Association participated in producing the brochures, and by 2001, did the groundwork for making certain that as many markets as possible were listed.

Totals for Available Years

Table 3.1 shows the number of markets from known published sources for Oregon. During the early part of the 20th century, it is probable that

TABLE 3.1. Inventory of Oregon farmers' markets, 1915 to 2005.

Year Represented	Number of Markets	Source
1915	3	Bettis (1969); Eigo and Engeman (2002)
1918	1	Grogan et al. (1919)
1946	2	Wann et al. (1948)
1976/1977	12	Linstrom (1978)
1981	12*	Thiemann (1983)
1994	18	Johnson and Bragg (1994)
1996	24	Johnson, Lewis, and Bragg (1996)
1998	38	ODA (1998)
1999	42	ODA (1999)
2000	46	ODA (2000)
2001	54	ODA/OFMA (2001)
2002	61	OFMA/ODA (2002)
2003	62	OFMA/ODA (2003)
2004	61	OFMA/ODA (2004)
2005	68	OFMA/ODA (2005)

*"Roughly a dozen..." (Thiemann, 1983).

only the most prominent markets were identified in national inventories; the national inventories did not include all of the major markets operating in Oregon. This issue perhaps improved during the inventory of the 1970s, but our most reliable information begins with the 1994 USDA inventory (Johnson & Bragg, 1994) and the ODA/OFMA directories. From 1994 on, the number of farmers' markets has grown steadily with a net gain of 50 markets during this period.

What Is Driving the Current Growth of Oregon's Farmers' Markets?

This section was stimulated in part by interests of the Project Advisory Committee. When the Committee began meeting to design the market manager survey, a discussion developed regarding why so many markets were forming. Discussed in detail next, the role of community building was an aspect of farmers' markets the Committee was interested in.

Table 3.1 shows the steady growth of farmers' markets in Oregon over recent decades. The 2002 survey of farmers' market managers reveals that markets ranged in age from 1 to 60 years (for the survey year). Most are relatively new. Table 3.2 indicates the general age categories and range of establishment dates for markets that participated in the 2002 manager survey. The data clearly show the growth in numbers of markets over the previous 10 years—68% were established during that period. Nearly one half (48%) of the markets were established in a 5-year period between 1996 and 2001. This reflects the rapid expansion in number of markets beginning in the 1990s and continuing to the present, and mimics similar growth in numbers of markets at the national level. What is driving this extraordinary growth?

Because the major link between farmers' markets and consumers historically has been food, a continued and increasing interest in the United States in high-quality food, or food not generally available through other venues, is likely to be one factor responsible for the increase in numbers of farmers' markets. Guptill and Wilkins (2002) noted that the current emphasis on local foods is the result of the "confluence of several movements: environmental, which gave rise to bioregionalism; community food security movement; and activism challenging the dominance of large corporations" (p. 39). Using a variety of sources, Hendrickson (2005) identified several important food trends in the decade between 1994 and 2004. These included increases in cooking schools from 338 to 930; viewers of the Food Network from 7 million to 79 million; the number of certified organic farms from 4,050 to 11,998; and, interestingly,

TABLE 3.2. Age categories of Oregon markets, 2002.

Age in 2002 (and Year Established)	Number of Markets	Percentage of Markets	Cumulative Percentage
1–5 years (2001 to 1996)	24	48	48
6–10 years (1995 to 1991)	10	20	68
11 or more years (1990 or earlier)	16	32	100
Total	50	100	–

the number of artisan cheeses participating in the American Cheese Society's annual competition, from about 100 in 1994 to more than 700 in 2004.

However, there is more than food driving the growth. This study identifies "community building" as a factor that fueled the growth of markets during the 1990s, and it continues to be a factor in their current growth. Before examining the data, it will be useful to revisit the decades leading up to the 1990s. As indicated previously, the period of growth of farmers' markets during the 1970s was stimulated by political action and Public Law 94-463 (Brown, 2001). Consumer interest in fresher food and the "back-to-nature" trend were also noted as other factors fueling farmers' market growth during that period. Studies from the 1980s identify fresh food as a major factor in the popularity of farmers' markets. A study of consumers at farmers' markets and supermarkets found that shoppers at farmers' markets were drawn there for the freshness of the produce, while shoppers at supermarkets cited convenience as the most important factor (Lockeretz, 1986). Other frequently cited consumer studies from that period have identified freshness and quality as leading reasons for shopping at farmers' markets (Adrian, 1982; Estes 1985; Pelsue, 1984).

However, there are other factors cited for the growth of farmers' markets in the late 1970s and early 1980s that are rarely associated with current growth of farmers' markets. For instance, access to less-expensive food is a theme for studies of farmers' markets from that period. Sommer, Wing, and Aitkens (1980) compared prices for food at Californian farmers' markets and supermarkets and concluded that farmers' markets are a bargain. A poor economy and high unemployment rates are also considered to be factors in the popularity of farmers' markets during that period. A study from Colorado cites an inflation rate of 13% as a reason why consumers are attracted to farmers' markets. In comparing supermarkets with farmers' markets, the authors calculate that consumers can save 29 cents per dollar on produce at farmers' markets. They conclude: "consumer demand during 1981 for these markets should be even greater because of continued inflation and the high unemployment rate" (Chase & Winn, 1981, p. 13). Moreover, in Oregon, one publication associates the origins of a farmers' market with the need to provide "local employment to offset the depressed economy resulting from lumber industry layoffs" (Thiemann & Burt, 1983, p. 3). Cheap food and a poor economy are rarely cited as primary reasons for the current growth in farmers' markets. This indicates that many of the markets that formed during the late 1970s and early 1980s did so under different circumstances, and were thus formed for different reasons than markets formed during 1990s and the current era. As we will see, some of the earlier markets were formed exclusively as a marketing channel for farmers.

Brown (2001) identified the 1990s as one of four growth surges for farmers' markets but because claims by other researchers for the growth surge are unsubstantiated she was unable to explain its occurrence. The literature tends to focus on the benefits farmers' markets provide, but not on what is driving the growth. For instance, one study stated that "farmers' markets are beneficial to farmers, consumers, local communities and the environment" (Baber & Frongillo, 2003, p. 87). These benefits are generally accepted among farmers' market advocates, but they do not tell us why so many farmers' markets have been established since 1999.

GETTING A HANDLE ON "COMMUNITY BUILDING"

Farmers' markets serve a role in Lyson's (2004) concept of "civic agriculture." "As social institutions and social organizations, farmers' markets can be important components of civic agriculture. They embody what is unique and special about local communities and help to differentiate one community from another" (p. 93). Lyson (2004) goes on to say that activities related to locally based agriculture "are tightly linked to a community's social and economic development" (p. 1). Simply put, this identifies the community-building functions of farmers' markets. These functions include the value of drawing high numbers of people together to interact; establishing a community identity; and, in some locations, shopping and interacting in businesses associated with the farmers' market site. In this way, farmers' markets help to form a bond among community members and strengthen local economies. The quality of community life is enhanced through these activities.

Why Oregon's Farmers' Markets Formed

As noted previously, while the Project Advisory Committee was assisting in developing the survey questionnaire, the committee members commented that they felt there had been a change, over time, in the reasons why markets were formed. All the committee members had recently been involved in organizing new farmers' markets. They pointed out differences between older and more recently formed markets. A major difference was in the original function of the market: a marketing channel for farmers or a catalyst for building communities. They devised a question that posed three alternative reasons for why markets were originally formed. The two main alternatives were "farmer income" and "community building." A third alternative was a "blend of community building and farmer income." The third alternative may be seen as the easy way out for survey participants, but, as one Committee member commented, "In 1991, I was part of a group that worked very hard to develop a market that would address both needs."

Table 3.3 shows the number and percentage of markets that were originally formed with the primary goal of farmer income, community building, or a blend of community building and farmer income. About one third of the markets fall into each category. However, as indicated previously, some older markets were formed under different circumstances than more recent markets. Is there a relationship between when markets were formed and their primary goal?

Statistical analysis revealed a significant relationship between the age of a market and why it was originally formed.[2] Additional analysis revealed that "community building" and "blend of community building and farmer income" appeared as goals during the same time period—a period of time different from the "farmer income" category. To explore this information in more detail, three age categories were created based on frequency distributions for the age of the market and the original reason for its organization. They are presented in Table 3.4.

TABLE 3.3. Why individual farmers' markets were originally formed.

Reason Market Formed	Number of Markets	Percentage of Markets
Farmer income	16	32
Community building	17	34
Blend of community building and farm income	17	34
Total	50	100

TABLE 3.4. Periods when markets were organized and the primary reason for organizing.

Period	Farmer Income	Community Building	Blend of Community Building and Farmer Income	Total
1986 and earlier	8 (66.7%)	2 (16.7%)	2 (16.7%)	12
1987–1994	6 (46.2%)	3 (23.1%)	4 (30.8%)	13
1995–2001	2 (8.0%)	12 (48%)	11 (44.0%)	25
Total	16	17	17	50

Although a small portion of markets organized during the period of 1986 and earlier were organized with community building in mind, the majority (about 67%) were organized primarily for farmer income. During the period 1987 to 1994, the appearance of community building as at least part of the original reason for forming the market begins to become significant. Nearly 54% of markets cited community building or a blend or community building and farmer income as primary goals. Furthermore, it is striking that community building dominates the period 1995 to 2001, with 48% of markets formed during this period citing the priority of community building, and a total of 92% formed with either community building or a blend of community building and farmer income as goals. Only 8% of markets were organized primarily for farmer income during that period. There is a fundamental change in thinking between the farmers' market organizers of the late 1970s/early 1980s and the 1990s.

The contemporary use of markets to feed, as well as to bring together communities, is apparent in the numbers of markets organized for these purposes between 1995 and 2001. A number of Oregon farmers' markets are based primarily on a model of enhancing their communities by offering high-quality food and a community identity. One market, the Hollywood Farmers' Market, is most notable for this. Organized in 1997 in one of Portland's many residential districts, it was intended to enhance a sense of community in an area through which people tended to drive without stopping. Even the neighborhood residents left to obtain most of their goods. Organized by just six community members, this market operates on a philosophy of high community involvement and volunteerism. This effective model has been duplicated at other farmers' markets.

Continued Adaptation

Evidence of the ability of markets to adapt through modification of their original goals is of interest. As a follow-up question as to why markets were originally organized, the market manager survey posed the question of what the current goal of their market was. It used the same categories for the answer options: "farmer income," "community building," "blend of community building and farmer income." Table 3.5 shows that most markets have modified their goals.

According to managers, markets that formed originally for farmer income are incorporating community building as a goal, and markets that formed originally for community building are incorporating farmer income as a goal. Currently, 70% of markets have a blend of community building and farmer income as the primary goal. This is up from 34% of markets that cited a blend of these as their primary goal when originally formed.

Table 3.6 shows that the change has occurred regardless of when the market was organized. Virtually all markets have made the transition toward having a blend of community building and farmer income as their goal. A single market made the transition from having a community-building focus to a primarily farmer-income focus.

TABLE 3.5. Current goal of individual markets.

Goal	Number of Markets	Percent of Markets
Farmer income	7	14
Community building	8	16
Blend of community building and farmer income	35	70
Total	50	100

TABLE 3.6. Periods when markets were organized and their current goal.

Period	Farmer Income	Community Building	Blend of Community Building and Farmer Income	Total
1986 and earlier	3 (25%)	0 (0%)	9 (75%)	12
1987–1994	4 (30.7%)	1 (7.7%)	8 (61.5%)	13
1995–2001	1 (4.0%)	7 (28%)	17 (68.0%)	25
Total	8	8	34	50

The reason for this shift in goals among so many markets is not completely understood, but is likely to be related to the key characteristics of good farmers' markets discussed in chapter 6 and the need for markets to maintain a supply of high-quality products along with links to their communities. As Hinrichs (2000) concluded in her examination of the economic and non-economic functions of farmers' markets, "Social ties, personal connections, and community goodwill are often appropriately seasoned by self-interest and a clear view of price" (p. 301).

Endnotes

1. Provisions of this law continue via the 2002 Farm Bill (Public Law 107-171) as the "Farmers' Market Promotion Program."
2. An ANOVA test for age of markets and the three categories for why the market was originally formed was significant ($F(2, 47) = 7.451$, $p < 0.01$). A Tukey HSD test revealed significant differences between the category "farmer income" and the other two categories ($p < 0.01$ and $p < 0.01$). There was not a significant difference between "community building" and "blend of community building" ($p = 0.967$). An examination of the frequency distribution revealed that the lack of difference between the two categories resulted from their appearance during the same time period, rendering them essentially the same.

CHAPTER 4

CONNECTING FARMERS' MARKET SIZE AND MANAGEMENT ORGANIZATION

In their popular overview of the history and value of public markets, Spitzer and Baum (1995) pointed out that "Management generally is recognized as the critical determinant for achieving long run success with any public market" (p. 67). A study of the organization of small farmers' markets in Kansas concluded that there is "a strong need for organization within a market to ensure vendor satisfaction" (Hughes & Mattson, 1992, p. 5). Despite the obvious importance of management as a factor in successful farmers' markets, there have been very few research studies on the topic.

An overview of farmers' markets in Pennsylvania identifies four types of organizational structures:

1. There is no organization among producers; farmers simply show up.
2. An arrangement is made with the owner of a private business; producers are tenants and the owner makes all the rules.

3. There is an unofficial agreement among producers; the market is organized informally, but collects fees and uses guidelines. Producers are members of an "association" that is not legally organized or incorporated.
4. There is an official agreement among producers; the organization is a legal entity with secure legal and tax status (Center for Rural Pennsylvania, 2002).

Most farmers' markets in Oregon would be classified as the structures described in numbers 3 and 4.

Central to the organization of a farmers' market is the role of the manager. Neil Hamilton's (2002) national overview of the legal aspects of farmers' market management noted: "A key ingredient in the operation of a market…is the person who runs the market on a day-to-day basis. This job is the responsibility of the 'market manager,' a person designated by the market organizers to be responsible for making the operational decisions" (p. 8). Some key tasks are required for a manager to successfully operate a public market:

- **Operations:** Supervising other staff; managing operating budget, including accounts receivable (rent) and payable; enforcing operating rules and procedures; maintaining relations with adjacent business properties.
- **Leasing:** Renting space to vendors.
- **Tenant assistance:** Assisting merchants with design and merchandising.
- **Marketing promotions and public relations:** Making a calendar of events; issuing press releases; organizing fund-raising events; maintaining relationships with community agencies, organizations, and local government.
- **Capital improvements**.
- **Long-range planning:** Assisting organizers in long-range planning based on knowledge on-site.

- **Reporting:** Providing regular reports to organizers regarding all areas above and proposes policy changes. (Spitzer & Baum, 1995, pp. 69–70)

In addition, and relation, to the aforementioned list, Oregon farmers' market managers generated a short list of major and mundane tasks that they regularly perform during their markets:

- Market setup and dismantling
- Putting out signage
- Fee collection
- Publicity and public relations
- Equipment inventory and maintenance
- Staffing the market information booth
- Directing traffic
- Vendor recruitment
- Bookkeeping
- Obtaining sponsorship and organizing fundraisers
- Tax preparation
- Market safety
- Special event coordination

The use of a manager can be associated with the size of a market. According to Hamilton (2002), "If the farmers' market is a relatively small and informal one (such as six farmers in the church parking lot one afternoon a week), the market manager's job is not large. On the other hand, managing a farmers' market with hundreds of vendors and thousands of customers may be a full-time responsibility" (p. 8). The number and intensity of the tasks outlined previously will likely be related to the size of the market as well.

Along with the value of market organization and the important role of a manager, researchers see market rules as being at the heart of a well-managed market. Again, Hamilton (2002) pointed out the necessity of having rules for governing the market: "It might be possible to operate a farmers' market without any form of written rules or policies for the vendors, but it would not be long before a market would face difficult issues...The reason farmers' markets use rules and policies is to prevent problems...from arising" (p. 9).

Market rules have been related to the size of a farmers' market. As a Florida study recommends, "The rules should conform to the situation at hand. Thus, the rules at a small, rural market need not be as specific as those at the larger markets" (Zimet, Hewitt & Henry, 1986, p. 295). Furthermore, "the degree to which there exist formal rules and regulations that govern the producer participants in a farmer retail vegetable market depends on the size of the market. Without good management/organization larger markets would fail" (Zimet et al., 1986, p. 291).

A Summary of Key Findings

This chapter includes a large amount of detailed information. In order to facilitate a more complete understanding of the data and the analysis process, the key findings are briefly summarized. Greater detail is provided in the third segment of this chapter.

Four categories for market size were developed for the purpose of analyzing market management structures. The categories, discussed in detail in the following text, are *Micro*, *Small*, *Medium*, and *Large*.[1] In addition to assisting with the overall analysis of farmers' markets, these market size categories can be considered analogous to stages of organizational development and can represent what the future might look like for a growing market. For instance, a Small market that anticipates becoming a Medium market can plan for its growth based on what management tools current Medium-sized markets are using. In addition, it is important to have an organizational structure that is appropriate for the size of a market. A Small market that has an organizational structure similar to that of a Large market may waste resources. A Large market that has an organizational structure similar to that of a Small market, as one market manager stated, "will implode in a couple of years or sooner."

Here are some important findings concerning what management tools are used for markets of various sizes:

- The data indicate that markets add management structure and management complexity as they increase in size.
- Each market size category uses a specific array of management tools. The use of management tools changes between market size categories. Knowledge of these changes will help markets plan for maintaining efficiency as they grow.
- There is a critical transition point for markets that are "smaller" as they are becoming "larger" markets. Specifically, this point occurs between what this research has identified as Small and Medium markets.
- Micro and Small markets use more management tools as they grow. This adds more management structure with growth. Tools would include site management tools, such as maps to assign spaces, and market governance tools such as written rules and boards of directors.
- Because the management structure used by smaller-size markets is already in place for most Medium and Large markets, Medium and Large markets add "management complexity" as they increase in size. This complexity consists of planning for and acquiring a paid manager, additional employees and stable revenue to support these positions, as well as increased effort in the form of the number of hours worked both in season and out of season. The tasks performed by managers in these size categories are more complex; for example, budgeting and planning systems are more sophisticated.

Figure 4.1 helps illustrate this. Micro and Small markets add management structure as they grow. Medium and Large markets add management complexity as they grow. There is a critical transition point between Micro/Small markets and Medium/Large markets. Medium/ Large markets should have appropriate management structure in place,

FIGURE 4.1. Market size and increasing management structure and complexity.

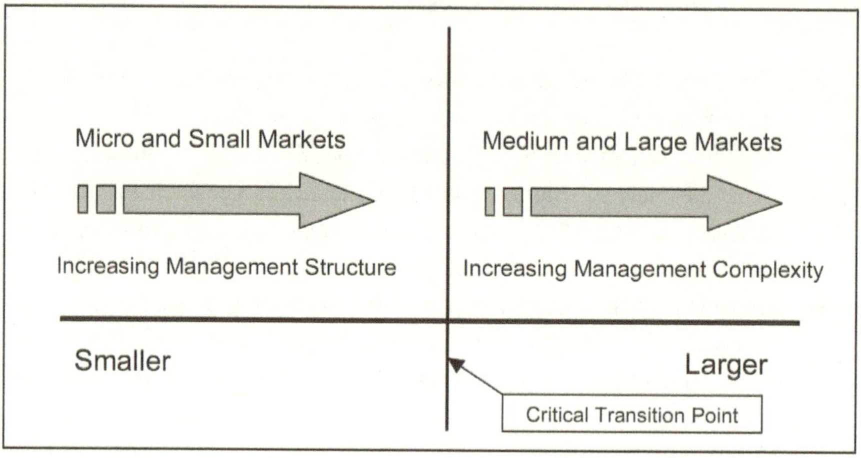

as well as the resources and planning for dealing with higher demands for paid labor and more hours worked in and out of season.

Because this discussion addresses what is *common* to market management, which may not necessarily be the best manner of operating, the conditions described may not represent the ideal. However, the knowledge of what operating guidelines other markets have instituted can help guide strategic planning and resource allocation for new markets or for established markets confronting growth or other significant changes.

It is important to point out that problems do not occur simply because a market lacks a management structure to prevent or address problems. The management or organizational structure must be functional. For instance, having a board of directors does not make a market viable. Having an effective board of directors, however, can make a difference.

MARKET SIZE CATEGORIES

Market managers, vendors, board members, and researchers routinely refer to markets according to size categories. It is common during a

Table 4.1. Size categories of Oregon farmers' markets (total market).

Category	Number of Markets	Percentage of Markets
Micro (5–8 vendors)	8	16
Small (9–30 vendors)	20	40
Medium (31–55 vendors)	12	24
Large (56–90 vendors)	10	20
Total	50	100

conversation for a market to be referred to as "small," "medium," or "large." Currently, there are no established definitions of market size, only general terms created to establish a framework for communicating about markets. Size categories are an important part of this research for comparative purposes, so considerable effort was directed toward creating workable categories. The categories were created using quantitative and qualitative data for guidance. The Appendix reviews the process for developing the categories used here. The categories represent the total size of the market, including both farmers and craft vendors, if present. Table 4.1 presents the size categories Micro, Small, Medium, and Large, along with the number of vendors associated with each category, and the number and percentage of markets in each category. These size categories are intended to be guidelines. The boundaries between categories should be seen as transitions rather than as hard divisions.

Examination of market size categories is largely missing from the research literature. The few studies that have used categories have tended to use convenient numerical breaks rather than identifying boundaries based on quantitative and qualitative information. A national survey of farmers' market managers conducted by the USDA in 1994 classified market size as 1–9, 10–19, 20–49, 50–99, 100–500, and more than 500 farmers (Burns & Johnson, 1996). Oberholtzer and Grow (2003) surveyed 57 farmers' market managers in the mid-Atlantic region of the United States (southeast Pennsylvania, Maryland, and Washington, D.C.).

The size categories they used for their analysis, along with the percentage of markets falling into each category, are as follows:

- 1–5 farmers (18.6%)
- 6–10 farmers (53.4%)
- 11–20 farmers (18.6%)
- 21–40 farmers (4.6%)
- 41 and more farmers (4.6%) (adapted from material on page 18)

These categories represent the number of farmers. A national survey of farmers' markets conducted by the USDA in 2000 (Payne, 2002) used these size categories: less than 10, 10–25, 26–50, and more than 50. These categories are fairly close to those used in this study and allow some comparison of Oregon to the national sample in the data analysis that follows.

MARKET MANAGEMENT ORGANIZATION

The manager survey used for this study included variables related to market structure and management complexity. The Project Advisory Committee generated a number of management methods or tools they felt would be associated with markets of different sizes. Other management tools, such as those related to manager effort and use of employees, were revealed during the data analysis. Following is a list of the tools used as variables in this analysis. They are only a small portion of the manager's tool kit, but they reveal important differences among markets which are related to size. The management tools cover several dimensions of market management. They are grouped as site management tools, market governance tools, and labor resource tools.

- **Site management tools:**
  - Manager designs market layout.
  - The position of vendors' spaces are arranged to meet specific goals.
  - Vendor spaces are assigned using a detailed map of the market site.

- **Market governance tools:**
    - The market has written guidelines.
    - The market has bylaws.
    - The market has a board of directors.
    - The board of directors has officers.
    - The board of directors has committees.
- **Labor resources tools:**
    - Manager compensation; managers are paid or are volunteers.
    - The market uses additional employees.
    - The number of hours the manager worked in season
    - The number of hours the manager worked off season

Each management tool is discussed and examined here in relation to the market size categories.

Site Management Tools

Many markets use a specific layout (bowling alley, horse shoe, circular, and so forth) to arrange vendors. A plan for market layout can be as simple as a mental visualization or, when coupled with a site map as discussed here, it can be fairly detailed. Markets differ in their need to have a specific layout. Having few vendors, some Micro and Small markets arrange themselves very informally. As markets increase in size, specifically designed layouts for markets are advantageous to enhance the flow of traffic and for other reasons. Although smaller markets vary as to whether their managers design a layout, the managers of all Medium and Large markets use a specific layout (Table 4.2). The decision concerning whether or not a layout is used for very small markets may reflect a manager's personality as much as it does the desire to manage the market efficiently. A large market lacking a planned layout would be chaotic and would quickly create discord among vendors and confusion among customers.

In addition to a layout plan to organize the market, managers arrange vendors to meet aesthetic and social goals. According to market

TABLE 4.2. Site management tools and market size.

Management Tool	Micro (5–8) $n = 8$	Small (9–30) $n = 20$	Medium (31–55) $n = 12$	Large (56–90) $n = 10$
Manager designs market layout*	75%	75%	100%	100%
Arrange vendors to meet goals**	50%	80%	100%	90%
Spaces assigned using a site map***	38%	65%	100%	100%

*$t(48) = -4.2245$, $p < 0.0001$. **$t(48) = -2.014$, $p < 0.05$. ***$t(48) = -5.873$, $p < 0.0001$.

managers, these goals include keeping agricultural products visible at market entrances, enhancing traffic flow, keeping the market visually interesting, and reducing friction between specific vendors. Again, Micro and Small markets are mixed as to whether they use or need to use this management tool, whereas Medium and Large markets virtually always use this tool (Table 4.2).

Another tool used by managers is a detailed site map with vendor spaces delineated. Prior to market day, a site map is used to assign specific vendors to specific spaces. Site maps correspond to markings identifying the spaces, which are created with chalk, traffic pylons, and paint at the market site on market day. Markets differ in their use of site maps according to the size of the market. Again, Micro and Small markets may or may not use a detailed map to assign spaces, but all Medium and Large markets do (Table 4.2).

A few other studies have investigated how managers organize the market site. In a Kansas study of small farmers' markets (Hughes & Mattson, 1992), it was noted that of the eight markets that had some type of formalized organization, only two assigned stalls on a seasonal basis. All the other markets assigned stalls on a first-come first-served basis. A survey of New Jersey farmers' market managers (Govindasamy et al., 1998) revealed the following:

> In order to assign the spaces available in the facility, different criteria were used. First come first served was the prevalent method

> and most markets tended to honor seniority...Some markets, however, assigned spaces according to the size of the farmers' operations or by lottery drawing. (p. 7)

With regard to positioning vendors to meet specific market goals, the same study indicated that 58% of managers did not consider identical produce when placing vendors side by side. Only one manager stated that farmers' personalities were taken into account when planning the market layout (Govindasamy et al., 1998). As described previously, the overwhelming majority of Oregon farmers' markets assign vendor spaces in advance and arrange vendors to meet specific market goals.

Market Governance Tools

With the exception of some of the smallest markets, almost all markets use written rules to assist in managing the market (Table 4.3). According to market managers, market rules govern the day-to-day operations of a market. They define what the market is trying to accomplish: how it appears, who sells, and what is sold. Rules also reflect government regulations regarding licensing, certification, and food handling. They are often written to provide checks on vendors who have attempted to circumvent established rules or because incidents have illuminated a weakness in previous rules.

Although the survey data did not reveal whether the complexity of market rules increase as markets increase in size, an interesting additional question for the survey would have been "How many pages are your market rules?" Some markets have only a few photocopied pages that are provided to vendors. Other markets have extensive sets of rules. During the focus-group sessions, some managers revealed the size of their market rules. For instance, according to a manager of one Small and two Medium markets, "I try to stay on one 11 × 17 sheet. We're at the limit. Margins and point size got smaller." In contrast, an urban farmers' market that manages three Medium and Large markets publishes a 52-page *Vendor Handbook* as a guide for its vendors.

TABLE 4.3. Market governance tools and market size.

Management Tool	Micro (5–8) $n = 8$	Small (9–30) $n = 20$	Medium (31–55) $n = 12$	Large (56–90) $n = 10$
Written rules	63%	90%	100%	90%
Bylaws*	25%	50%	83%	80%
Board of directors**	25%	65%	92%	90%
Board of directors has officers***	0%	55%	92%	90%
Board of directors has committees****	0%	25%	59%	70%

*$t(48) = -2.897$, $p < 0.01$. **$t(48) = -3.745$, $p < 0.001$. ***$t(34) = -6.302$, $p < 0.0001$. The values for markets that do not have boards of directors were not used in the statistical calculation. ****$t(35) = -3.268$, $p < 0.005$. The values for markets that do not have boards of directors were not used in the statistical calculation.

Most markets have written rules of some type, but some have bylaws. Bylaws are generally associated with formal governance arrangements such as having a board of directors or being defined as a corporation. Bylaws set out the "rules" for how board members are selected, how officers are elected, and so forth. As markets grow in size, bylaws become more common (Table 4.3).

Boards of directors are another management structure strongly associated with the size of a market. As markets grow, they add boards of directors. Table 4.3 again shows that Micro and Small markets have a mixture of approaches, while Medium and Large markets generally have boards of directors. The two larger markets (one Medium and one Large) that did not have boards of directors were both organized and/or operated by government or community entities (city government or downtown association). According to market managers, prior to having a formal board of directors, a market may use an informal advisory committee made up of vendors, or, as one manager commented, a manager may use a "kitchen cabinet" of a few hand-selected vendors to offer input.

Boards of directors span a continuum from informal to very formal. They simply may be an advisory committee using the term "board of directors," or they may be highly formalized boards that follow strict protocols. The role the board assumes differs from market to market. Some are very "hands on" or active, while others work exclusively in an advisory capacity.

Some boards of directors govern via egalitarian or consensus approaches. For instance, smaller markets may use the term "board of directors" even though they are not structured as a nonprofit organization requiring a board. As markets increase in size, they elect officers to manage specific tasks or to provide leadership for a component of market management such as overseeing its financial condition. Of the 35 markets that have boards of directors, only the smaller markets do not necessarily have officers as part of their board structure while most larger markets do have officers (Table 4.3). Boards of directors and bylaws are required for markets that have the legal tax status of a "not-for-profit" organization in Oregon. Many markets, however, use these tools even though they do not have this status.

Very few studies have documented the use of boards of directors by farmers' markets. Payne (2002) reported that 63% of farmers' markets from a national sample use "vendor operated" boards (not defined) (Table 4.4). It is surprising to note that although the Oregon results show significant differences in the use of boards of directors by markets of different sizes, Payne's data show that the use of boards is consistent across market sizes. For Oregon, 70% of markets use boards of directors. However, the use of boards varies widely according to market size, with smaller markets less likely to use one and larger markets more likely.

TABLE 4.4. Results of national survey: market size and vendor boards.

Market Size Category	Use Vendor-Operated Board
Less than 10 farmers	62%
10–25 farmers	64%
26–50 farmers	60%
More than 50 farmers	61%

Source. Payne (2002).

The addition of committees to boards adds another layer of organizational structure. Committees may be standing or ad hoc and generally indicate that there is a level of complexity in managing a market that requires tasks to be delegated to segments of the board. These tasks might include functions that were handled by a manager within the organization of a smaller market. The addition of committees to boards of directors is associated with larger markets (Table 4.3).

Labor Resource Tools

Manager Compensation and Market Size

A good starting point in the examination of management complexity as it relates to size of markets is whether or not the manager is compensated financially for her/his efforts. Fourteen markets (28%) use volunteer managers and 36 markets (72%) use paid managers. There is a clear relationship between the size of markets and the status of the manager. Table 4.5 shows the different sized markets managed by volunteer or paid managers. It is significant that there are no Medium or Large markets managed by volunteers. This may indicate that the management of markets of this size involves a level of complexity that requires more labor than volunteers are able or willing to commit. This relationship between the size of the market and volunteer management is noteworthy. Chapter 5 includes a discussion regarding the potential threshold for numbers of hours committed for volunteer managers.

A survey of New Jersey farmers' market managers (Govindasamy et al., 1998) revealed that about 29% of managers are volunteers. The authors do not describe the scale of the markets in this study, nor do they associate market scale with any of the other variables they examined. Managers of farmers' markets in the mid-Atlantic region (Oberholtzer & Grow, 2003) revealed that about 44% are volunteer managers. Again, the use of volunteer or paid market managers was not examined in relation to market scale. The Kansas study of small farmers' markets (Hughes & Mattson, 1992) indicated that five out of the eight coordinators of organized markets were paid.

TABLE 4.5. Labor resource tools and market size.

Management Tool	Micro (5–8) $n = 8$	Small (9–30) $n = 20$	Medium (31–55) $n = 12$	Large (56–90) $n = 10$
Manager is a volunteer*	55%	47%	0%	0%
Manager is paid*	45%	53%	100%	100%
Market uses additional employees**	0%	20%	92%	100%
Additional employee FTE per market (mean)***	0	0.10	0.54	0.80
Hours per week in-season per market (median)****	7.0	10.0	20.0	29.0
Hours per week off-season per market (median)*****	2.0	3.0	10.5	15.5

*$t(48) = -4.917$, $p < 0.0001$. **$t(48) = -8.628$, $p < 0.0001$. ***$r(50) = 0.773$, $p < 0.0001$. ****$r(47) = 0.705$, $p < 0.0001$. *****$r(47) = 0.608$, $p < 0.0001$.

Payne's (2002) national survey of farmers' markets reported that 30% of all markets use paid managers. Of the markets with paid managers, 51% were full-time (15% of the total sample) and 49% were part-time (14% of the total sample). Payne (2002) also associated market size categories with paid managers and with full- or part-time paid managers (Table 4.6). His data show a correlation between paid managers and markets of larger size; only 18% of markets with fewer than 10 farmers use a paid manager, while 71% of markets with more than 50 farmers use a paid manager. In contrast, 28% of Oregon's markets use a volunteer manager and 72% use a paid manager. All Oregon markets with more than 30 vendors use a paid manager.

Markets use employees to assist in managing a number of general and highly specialized tasks. Markets differ in their use of additional employees, and there is a strong association between the use of employees and the size of the market. Micro and Small markets rarely used additional paid employees (Table 4.5). Remember that volunteers manage many of these markets, and it would be unlikely for paid employees to be used under those circumstances. Conversely, virtually all Medium and Large markets used additional paid employees.

Not only do markets use additional paid employees as they increase in size, the number of employees and/or the number of hours they work

TABLE 4.6. National survey results: market size and full- and part-time managers.

Market Size	Part-Time	Full-Time	Total
Less than 10 farmers	10%	8%	18%
10–25 farmers	10%	14%	24%
26–50 farmers	18%	20%	38%
More than 50 farmers	46%	25%	71%

Source. Payne (2002).

increases as the markets increase in size. Table 4.5 shows that as markets increase in size, the total full-time equivalent (FTE) for all employees increases during the market season.

Market Size and Manager Effort

The number of hours managers work varies by the number of markets they mange, whether it is the market season or the off-season, and the size of the market(s) they manage. As a market increases in size, its management demands more hours of work. Table 4.5 shows the median number of hours worked per manager per market during the market season for the different market-size categories.[2] Managers of Micro markets work a median of 7 hours per week during the market season, ranging from 1 to 15 hours per week. Managers of Small markets work a median of 10 hours per week during the market season, ranging from 2 to 30 hours per week. Managers of Medium markets work a median of 20 hours per week, ranging from 8 to 25 hours per week. Managers of Large markets work a median of 29 hours per weeks, ranging from 16 to 40 hours per week.

As noted previously, the number of hours worked during the off-season is influenced by market size as well. Table 4.5 clearly shows the progression in number of hours worked per manager per market during the off-season as the market size increases. Managers of Micro markets work a median of two hours per week during the off-season, ranging from 0 to 15 hours per week. Eighty-three percent of these managers work four or fewer hours per week during the off-season. Managers of Small markets work a median of three hours per week during the off-season, ranging from 0 to 15 hours per week. Managers of Medium markets work a median of 10.5 hours per week, ranging from 0 to 25 hours per week, with nearly 67% working 11 or fewer hours per week. Managers of Large markets work a median 15.5 hours per week during the off-season, ranging from 7 to 40 hours per week. In summary, Table 4.5 clearly shows the increase in management effort, defined as hours worked both during the market season and during the off-season, as markets increase in size.

When farmers' market managers were shown these results during focus group sessions and during the Project Advisory Committee review session, they agreed and noted other aspects of increasing management complexity. Their comments help flesh out some of the issues associated with the growth of a market. They noted increasing demands during the off-season:

- "During the off-season, the kind of work changes too with the size, not just the hours. Involvement with bureaucracies, insurance, et cetera, not just taking calls from vendors."
- "I'm sending out vendor packets earlier because I couldn't handle the paperwork load last year."
- "I prioritize off-season tasks—no musicians booked in January!"
- "The off-season is getting shorter; market is opening earlier, staying open longer, then there are special end of season and holiday markets. And, there's more to do during the off-season related to regulations. Plus negotiating and renewing insurance."

With increasing size comes an increase in product diversity and the demands for additional skill and responsibilities:

- "The breadth of knowledge is more too. Food safety [Oregon Department of Agriculture, Food Safety Division] wants managers to know and enforce, vendors want farm inspections. A lot of regulation is put on managers."

- "And we need to know as we add meat, kitchens, restaurants, what kind of information to give the vendors."

There is also the difference in organizational complexity and management effort as markets become large:

- "About 50 vendors is the perfect size. A market much larger than about 50 vendors is three times more difficult to manage."
- "With large markets internal controls are a necessary piece of infrastructure. Financial issues are complicated."
- "For large markets there's more emphasis on mission and vision and strategic plan and goals and accountability. I now do more public relations and create an advertising plan. Budgets; my time has become more and more about planning, budgeting, etcetera."
- "There are some economies of scale but there's always something added to nullify them. They're there but they just get washed over by some other thing you're now expected to do."
- "Where my workload has changed over the last years is in finance, budgeting, accounting for budgets. A lot of time goes into that."

There are other qualitative differences between Medium and Large markets as well. Reflecting on the transition from a Medium market to a Large market, one manager commented:

> This got me thinking about the evolution from starting six years ago. When I started, the markets were smaller and growing into larger markets. What did we have to put into place to make that change happen? What's different between managing a medium sized market and jumping into something large? It's the size of course; it brings you more problems. The main thing is you have to let go of so much. You can't be everywhere and you don't know what's happening at that other end of the market now. You have to trust more, have more structure, and have systems for enforcement of rules. You can't go around and make sure everybody's doing everything every week. You have a system where you go down one leg one week, the next the next. You try to hit things

> consistently, but you can't do it all. The nature of the relationships changes with the vendors. When it's a smaller market you have a much closer relationship. And so vendor problems are part of it; you have more vendor problems with a large market. They don't 'mind.' If you are at one end of the market, and they can't see you, they try to sneak in and try to load out early, whatever they want to do. There's a lot more time spent as policeman instead of as a friend and a confidant kind of relationship.

Table 4.7 arranges the management tools in an order approximating the increase in management structure and complexity for the four market size categories. Based on these data, some characteristics for each market size category can be identified:

- Micro markets are very informal. They are the most likely size of market to have a volunteer manager. Because of the low number of vendors, there is not the demand to use a site map or to arrange vendors to meet specific goals. Many do not have written rules or guidelines. Only about seven hours per week are devoted to

TABLE 4.7. Summary of market size and management structures.

Management Tool	Micro (5–8)	Small (9–30)	Medium (31–55)	Large (56–90)
Manager is a volunteer	55%	47%	0%	0%
Manager is paid	45%	53%	100%	100%
Design layout	75%	75%	100%	100%
Guidelines	63%	90%	100%	90%
Site map	38%	65%	100%	100%
Arrange vendors	50%	80%	100%	90%
Board	25%	65%	92%	90%
Bylaws	25%	50%	83%	80%
Board officers	0%	55%	92%	90%
Board committees	0%	25%	59%	70%
Additional employees	0%	20%	92%	100%
Median hours per week in-season per market	7.0	10.0	20.0	29.0
Median hours per week off-season per market	2.0	3.0	10.5	15.5
Additional employee mean FTE per market	0	0.10	0.54	0.80

managing these markets during the market season. Virtually none use any of the organizational structures associated with larger markets, such as boards of directors, bylaws, and so forth.

- Small markets are more complex than Micro markets and show many signs of more formal organization. They are equally split between those that have paid managers and those that have volunteer managers. They are large enough to use rules or guidelines and arrange vendors to meet specific goals. They have substantially more vendors than Micro markets but use only a few additional hours each week for management during market season and off-season.
- The transition from a Small market to a Medium market is identified here as a critical point in a market's growth. There are significant differences in organizational structure between the two size categories that must be taken into account in order to successfully make the transition from a Small to a Medium market. Markets in this size transition should have a system in place for supporting a paid manager and additional part-time employee(s). A more formalized structure that includes a board of directors and bylaws, and more detailed market rules is necessary for success. There are significant increases in the number of hours required to manage a Medium market as compared with a Small market. The median hours required for management double during the market season and jump from 3 to almost 11 hours a week during the off-season.

As pointed out in the first segment of this chapter, there are many similarities between Medium and Large markets. The differences between these categories are those of increased complexity in the form of greater labor and resource requirements, rather than additional management tools. The number of hours worked by the market managers per week during the season and the off-season is higher for Large markets than for Medium markets. Both Medium and Large markets use additional employees, but all Large markets use additional employees, and they require a greater number of hours of work by the additional employees than do Medium markets. The comments of managers cited previously described some of the qualitative differences between these market size categories.

ENDNOTES

1. Throughout the text, these categories will be capitalized: Micro, Small, Medium, and Large.
2. There are seven managers in this study who manage more than one market. In order to examine the impact of market size on manager effort and to avoid problems associated with managers who manage multiple markets of differing sizes, the values used here are proportioned to reflect per-manager per-market effort. This does not take into account additional effort managers of multiple markets may devote to specific markets during startup or to dealing with a variety of issues or efficiencies of scale from managing more than one market.

Chapter 5

When Things Don't Work: Some Insights Into Why Farmers' Markets Close

> I have an urgent plea for help. [Our] farmers' market currently has no produce except berries, and has no farmers planning to bring produce until late July or (more likely) early August…I know there must be some farmers out there with lettuce and other vegetables. I'm getting desperate. We have customers asking every week "where are the vegetables?" and lots of folks in our community have the WIC and Senior coupons to spend at the market. It's very frustrating being in a farming community and not having any farmers at the market! (Farmers' market e-mail discussion list posting, July, 2002)

This appeal epitomizes a farmers' market manager struggling with a market in distress. Despite laudable efforts by its organizers, this market closed permanently a few months later. The satisfaction of observing the increase in number of farmers' markets in Oregon and elsewhere obscures the little-known fact that a disturbing number of them fail.

MARKET DYNAMICS

Economists associated with the U.S. Department of Agriculture (USDA) periodically inventory and analyze farmers' markets in the United States. Several such reports have been published during the period of growth of farmers' markets from the 1990s to the present (Burns & Johnson, 1996; Johnson & Bragg, 1994; Johnson et al., 1996; Payne, 2002). These reports have documented a significant expansion in the number of farmers' markets nationwide. Although they largely recorded farmers' market start-ups, the authors also recognized that there were other processes taking place. As Burns and Johnson (1996) noted, "Not all farmers' markets are successful and only anecdotal information exists on why some have failed" (p. 12). Later, Payne (2002) pointed out, "While farmers' markets have shown that they are beneficial to farmers, customers, and local communities, many areas of study remain. One of the most important areas that merits further study is why markets fail" (p. 9).

Market Failure

Chapter 3 reviewed periods of growth and decline of numbers of farmers' markets over the past 90 years. Within that discussion, a more accurate picture of the number of farmers' markets was presented using data from market directories which were enhanced by data from other sources. These directories were published by the Oregon Farmers' Market Association (OFMA) and the Oregon Department of Agriculture (ODA) beginning in 1998. Table 5.1 presents the number of farmers' markets in Oregon for the eight years available. It shows a largely steady increase in the number of markets from 38 in 1998 to 68 in 2005. This type of information is commonly presented by researchers for this period of growth in the number of farmers' markets. For instance, Thilmany and Watson (2004) used USDA data to document the growth of farmers' markets in the United States. They noted that the growth of farmers' markets in the western United States between 1994 and 2002 was 168%, outpacing the national rate of 79%.

Table 5.2 shows the number of new markets that opened each year and the number of markets that operated during a specific season(s) but did not reopen. These markets closed permanently sometime during or after their final market season. For example, between the end of the 1998 season and the beginning of the 1999 season, 11 new markets opened while 6 markets that had previously operated did not reopen. Over the entire period from 1998 to 2005, 62 new markets opened and 32 closed. This is a very different picture than that presented in Table 5.1. Remember, this relatively high rate of failure for new markets occurred during a period of overall growth in the number of farmers' markets—a time when the farmers' markets sector was portrayed as booming.

Focusing on the 32 markets that closed between 1998 and 2005, the overwhelming majority had short life spans (Table 5.3). Of the

TABLE 5.1. Growth in number of farmers' markets, 1998 to 2005.

	1998	1999	2000	2001	2002	2003	2004	2005	Gain
Number of farmers' markets	38	43	46	58	61	62	61	68	30

TABLE 5.2. Number of markets, by year, that opened or did not reopen.

Markets	1998	1999	2000	2001	2002	2003	2004	2005	Total
New markets	N/A	11	5	16	9	6	4	11	62
Markets not reopening*	N/A	6	2	4	6	5	5	4	32
Total number of markets	38	43	46	58	61	62	61	68	–

*Exact date of closure is unknown (during a season, after a season, or just prior to the next season).

TABLE 5.3. Life span of markets that closed, 1998 to 2005.

Number of Years Operating*	Number of Markets	Percent of Markets That Closed
1	15	46.8
2	5	15.6
3	4	12.5
4	6	18.8
More than 4 years**	2	6.2
Total	32	100

*For six markets, the exact number of years of operation was unknown. They are listed here as operating the number of years known. Four are recorded as 1 year, one is recorded as 3 years, and one is recorded as 4 years. **These 2 markets operated for 11 and 22 years, respectively.

32 markets that closed, 15 (nearly 47%) did so following their first season. Thirty of the 32 markets (94%) closed after operating 4 years or fewer.

The fact that so many markets do not successfully make it through one season is startling. First-year markets appear to have a high risk level, but two markets closed, even after 11 and 22 years respectively. This indicates a complexity to the issues surrounding the failure of individual markets.

High failure rates of new firms are often cited in business literature. A recent analysis demonstrated that only 50% of new firms survived for more than 4 years (Headd, 2003). This is, to some extent, consistent with the data for farmers' markets. To obtain a sense of how farmers' markets compare to business firms in this regard, a failure rate was calculated using the method common in business literature (such as Headd, 2003). Using 2001 as the base year—because it had the largest number of market start-ups—the fate of the 16 markets that started that year was followed for 4 years. During that period, eight markets (50%) failed. This is a very small sample but does at least offer an indication of the high failure rate of new markets.

Market Manager Turnover

In addition to the surprisingly high level of failure for farmers' markets, another aspect that emerged during the analysis was the high rate of turnover of market managers. Although it was not initially an area of focus in the manager survey, turnover of farmers' market managers emerged as an issue during follow-up phases of the research. Attempts to contact managers to participate in focus groups and follow-up phone calls and e-mail regarding research details revealed changes in management for many markets which prompted further examination.

Using data from the OFMA/ODA directories, the number of farmers' markets operating under a new manager is listed in Table 5.4, by year, for the period 1999 to 2005. These figures represent the number of markets that began a season with a change in manager from the previous season and do not include new managers of new markets. For each year, a significant number of markets, ranging from 11 to 19, changed managers. The total for the eight seasons is a disturbing 101 manager changes, for a weighted annual average 30% turnover rate for managers of existing markets.

Not all manager turnover is negative. A change may improve a market, or reflect a better opportunity for a manager. Clearly, however, the very high rate of turnover documented here cannot be entirely positive.

Table 5.4. Number of existing Oregon markets operating under a new manager.

Existing Markets	1999	2000	2001	2002	2003	2004	2005	Totals
New manager	14	11	13	17	15	19	12	101 manager changes
Returning manager	18	30	29	35	41	38	45	N/A
Percent manager turnover*	44%	27%	31%	33%	27%	33%	21%	Weighted Average 30%

*Rounded to whole numbers.

Even under the best conditions, a change in manager produces some stress for a market.

RESOURCES, SUCCESS, AND FAILURE

Organizers of farmers' markets make many highly individualized decisions that may contribute to the ultimate success or failure of their market. However, all markets must obtain and use revenue and personnel resources to provide an environment for sound decisions and long term sustainability. This is particularly crucial for the overwhelming majority of markets that depend on vendor fees for operation. Markets vary greatly for both revenue and personnel resources. Local supplies of customers and products play a role in what revenue may be generated by a farmers' market. Revenue generally provides labor for management of market operations in order to sustain the market by continuing the cycle of attracting adequate customers and farmers.

Revenue Resources

In most instances, the value of revenue generated by a farmers' market shapes the type and sophistication of its management organization. Nearly all markets surveyed (92%) collect stall fees from vendors. These fees are the primary source of revenue for markets. Other sources of revenue for some markets include annual membership fees, promotional items, fundraising events, sponsorships, donations, and grants. Figure 5.1 summarizes the most common revenue resources.

Haves and Have-Nots

When USDA researchers inventoried farmers' markets in the mid-1990s, 84% of markets indicated that their operations were "self-sustaining" (not defined) (Burns & Johnson, 1996). A later USDA study indicated that a similar number (81%) were "self supporting," which is defined as "market income sufficient to pay for all costs associated with operating the market" (Payne, 2002, p. 4). Neither "self-sustaining" nor "self supporting" captures the wide disparity in administrative revenue

Figure 5.1. Sources of revenue.

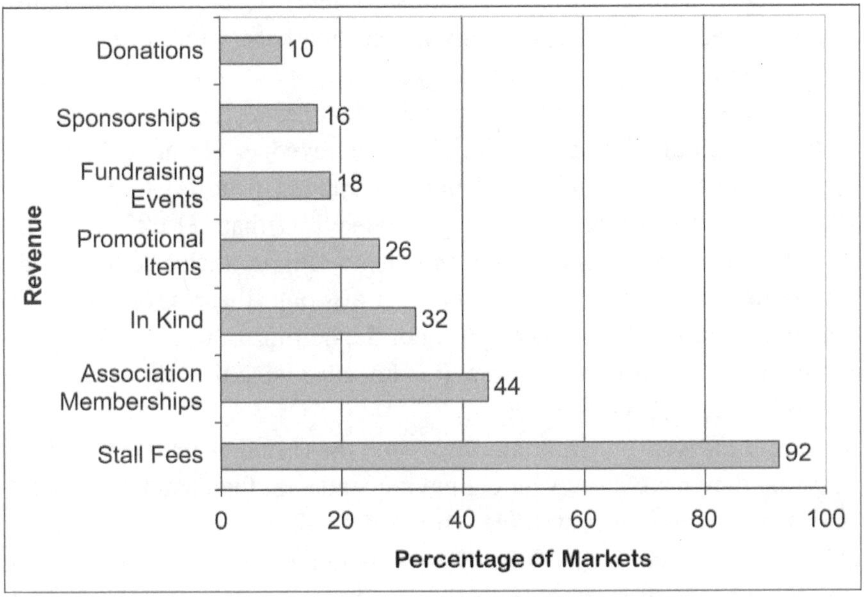

generated by farmers' markets. For most Oregon farmers' markets, "self supporting" means living within the confines of what is a very limited budget for operating a farmers' market.

Resembling the historic model of large indoor municipal markets, and analogous to contemporary shopping mall management, farmers' markets operate as an enterprise that provides organizational services to a group of independently operated enterprises (farmers, and in some instances, craft vendors). For these organizational services, the independently operated enterprises generally pay a "rent," which for farmers' markets is a stall fee. As noted previously, market administrations collect other forms of "revenue" too. For the 50 farmers' markets that participated in the manager survey, the sum of all sources of revenue received by market administrators (stall fees, sponsorships, etc.) totaled nearly $1 million for the 2001 season. To clarify, this sum is *administrative revenue*, the financial resources used for operating the market.

It is not revenue to vendors from sales (revenue to vendors is a substantially higher number and is not addressed here). This $1 million figure is distributed widely and unevenly among markets, with most markets operating on very low amounts of revenue.

Examining the 50 markets on an individual basis offers a clear picture of the uneven distribution of administrative revenue. The annual administrative revenue for individual markets ranged from $0 to $111,000 (Table 5.5). Six markets (12%) generated less than $1,000 each. In contrast, seven markets (14%) each generated more than $50,000, with one collecting more than $100,000. As a group, these seven markets accounted for nearly half ($482,641) of the administrative revenue generated by all 50 farmers' markets. It is not surprising that all seven were in the Large market-size category.

Larger markets generate more revenue than smaller markets for two reasons: they have more vendors paying stall fees than smaller markets and they charge higher stall fees ($r(50) = 0.502$, $p < 0.01$). To illustrate this, Table 5.6 shows that 80% of markets in the Large market-size category, and 75% of markets in the Medium market-size category, charge between $13 and $35 per stall. Conversely, 88% of markets in the Micro market-size category, and 65% of markets in the Small market-size category, charge between $0 and $12 per stall. Thus, smaller markets face two challenges to generating adequate market administrative revenue: fewer vendors from whom to collect fees and lower stall fees.

TABLE 5.5. Administrative revenue for individual Oregon markets.

Administrative Revenue	Number of Markets	Percent of Markets
$0–999	6	12
$1,000–4,999	12	24
$5,000–9,999	9	18
$10,000–19,999	8	16
$20,000–49,999	8	16
$50,000–99,999	6	12
$100,000 or more	1	2
Total	50	100

Table 5.6. Larger markets charge higher stall fees.

Market Size Category	*Category of Fees**				
	$0–8	$9–12	$13–20	$21–35	Total
Micro (5–8)	50% (4)	38% (3)	13% (1)	0% (0)	100%** (8)
Small (9–30)	35% (7)	30% (6)	25% (5)	10% (2)	100% (20)
Medium (31–55)	8% (1)	17% (2)	42% (5)	33% (4)	100% (12)
Large (56–90)	0% (0)	20% (2)	50% (5)	30% (3)	100% (10)
Total Markets	12	13	16	9	50

*Fee categories are arranged to approximate quartiles. **Percentages are rounded to nearest whole number.

To summarize, administrative revenue figures highlight the wide differences related to scale and location between farmers' markets. Of the nearly $1 million in annual administrative revenue generated by Oregon farmers' markets (2001 season), 7 of the 50 markets accounted for nearly half. Smaller markets generate less administrative revenue, and because of the connection between market scale and population centers, markets in more sparsely populated regions of Oregon have a smaller revenue resource to work with. It is important to remember that the philosophy behind farmers' markets is not necessarily to maximize market administration revenue. For the many markets that generate lower revenues, the value of factors such as the service they provide to customers, the income provided to the vendors, the improved nutrition and food security, and the enhanced sense of community are immense. That said, the viability of most markets is tied to their revenue stream. Administrative revenue level is connected to the access and use of important resources. The major enhancement that can be provided by adequate administrative revenue is paid labor to handle overall market operations.

Management and Labor Resources

All farmers' markets must handle both simple and highly complex tasks in order to operate during the market season and to maintain management operations during the off-season. All Oregon farmers' markets have someone who functions in the role of manager. Farmers' markets with sufficient administrative revenue hire part- or full-time managers, and sometimes additional paid staff. Those markets with lower administrative revenue use a volunteer manager and other volunteer labor. Manger positions span a continuum from full-time annual positions, to part-time seasonal positions, to positions that function only during the market season and on market day. Whether a manager is paid, as well as the level of pay, are linked to how much administrative revenue a market generates.

Figure 5.2 summarizes the major sources of labor that markets use in order to operate. Most markets (36) use a paid manager, but a significant number (14) use volunteer managers. Additional labor is supplied by volunteer staff and paid employees (in addition to the manager). Each of

FIGURE 5.2. Sources of farmers' market labor: managers and employees.

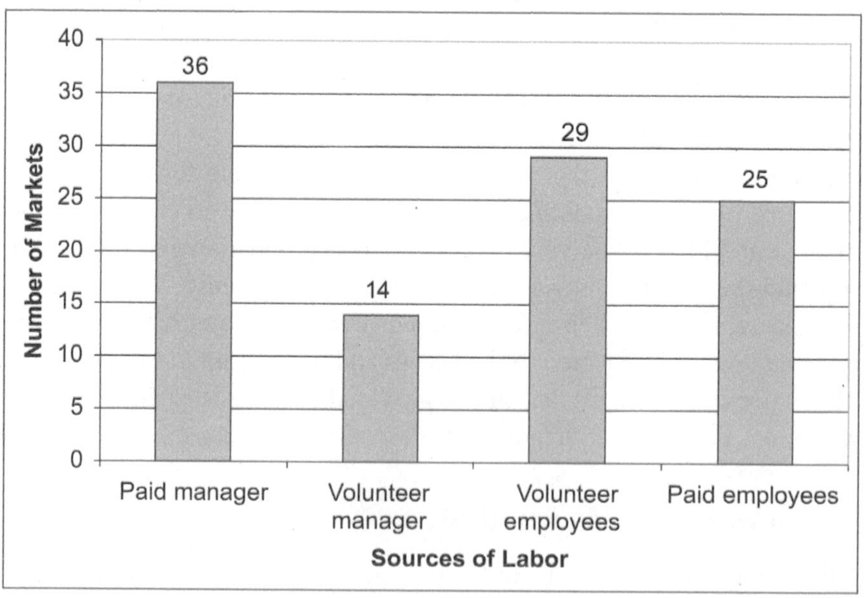

these sources of labor is discussed as follows. Keep in mind that some managers oversee more than one market.

Table 5.7 shows categories of administrative revenue for farmers' markets, with corresponding salary ranges for managers, in order to illustrate two points: volunteer managers are associated with markets that generate lower revenue and, more broadly speaking, the manager's salary level is correlated with the administrative revenue a market generates; markets with higher or lower administrative revenue compensate managers at higher or lower levels.

A national study of farmers' markets provides some indirect comparison to the link between revenue and the use of volunteer or paid managers. Using total market sales (total revenue to vendors from sales), Payne (2002) observed a connection between market sales (not administrative revenue) and whether markets had a paid manager. When markets had $10,000 or less in total sales, only 11% had paid managers. In contrast, when markets had more than $500,000 in total sales, 75% had paid managers.

The term "paid manager" is relative. The range of managers' salaries in 2002 was $650 to $35,000. Four managers made less than $1,500, placing them close to volunteer status. Four managers making $20,000

TABLE 5.7. Administrative revenue and salaries of Oregon managers, 2002.

Market Administrative Revenue	Salary Range of Managers	Number of Managers*
Less than $5,000	$650–$2,600	
	+ 11 volunteer managers	14
$5,000–$9,999	$1,000–$7,800	
	+ 3 volunteer managers	5
$10,000–$19,999	$2,700–$12,000	5
$20,000–$49,999	$10,000–$20,000	3
$50,000–$74,999	$9,000–$28,000	4
$75,000 or more	$23,000–$35,000	3

*These figures are for farmers' market organizations and include organizations that manage multiple markets. There are a total of 40 individuals managing 50 farmer's markets. Six managers are compensated by government or civic entities and are not included here.

to $35,000 approach what might be considered a living wage level of compensation (for 2002). To some degree, the compensation reflects the number of hours a manager works, as smaller, less management-intensive markets tend to require less effort. There are exceptions to this. Of concern to this analysis are situations in which a high level of effort is required to manage a market but the market administrative revenue is insufficient to pay an adequate salary.

In sum, the amount of administrative revenue and the labor resources a farmers' market can provide are closely tied to its size. Smaller markets may experience a circular condition: These markets have a relatively small number of vendors, and, as a result, the amount of administrative revenue from stall fees is relatively low. The amount of revenue available affects whether the market can employ a manager and how much the manager can be paid. This, in turn, affects how much time and effort can be expended managing the market, which includes recruiting vendors and customers. This circular condition easily turns into a downward spiral. This situation is explored further in the next section.

A Few Insights Into Markets That Close

Presented in this section is what can be gleaned from a relatively small amount of data about farmers' markets that have closed. The data are from three sources: the 2002 survey of farmers' market managers, the OFMA/ODA farmers' market directories for 1998 to 2005, and interviews with managers of closed markets and other farmers' market managers. Nine markets that were operating at the time of the survey of market managers closed sometime between when the survey was conducted and 2005. (To clarify, Table 5.2 shows a total of 10 markets closing during 2003 and 2004. One of these markets opened in 2002 and was not eligible to participate in the survey. Therefore, nine markets are included here.) The 2002 survey of market managers includes data on seven of these nine farmers' markets (two markets did not participate in the survey). The OFMA/ODA farmers' market directories provide limited but important information on all nine of the closed markets. Again, this is only a small sample of markets but the analysis still sheds some light on the phenomenon of market closures.

The Market Dynamics section analyzed market failure and manager turnover on an aggregate basis. Here the focus is on the nine markets that closed after the 2002 survey. Presented here is some background on the nine markets as a group, followed by five factors associated with their failure, comments on the process of failure, and some additional concerns regarding market failure.

Eight of the nine markets that closed were community-based markets operated as grassroots organizations. The ninth market was sponsored by a business that provided space for vendors but did not collect fees or provide direct management. Two of the community-based markets were associated with market organizations that managed more than one market. In these instances, the markets closed but the market organizations continued operating and managing other markets. One market was in a highly urban setting, four were in rural settings, and the remaining four were in communities with sufficient populations to support a market in general and perhaps even one of moderate size. All of the markets were located within a reasonable distance from product supply (farmers).

Five Factors Associated With Farmers' Markets That Fail

Factor 1: Small Size

All nine of the markets that closed were in the Micro or Small size categories, with five markets in the Micro size category and four markets in the Small size category. Survey data indicate that the number of vendors for total market size (including craft vendors) ranged from 5 to 20, and that the number of farmer vendors (source of food products) was between 4 and 13. Because the survey was conducted during 2002 (based on vendor numbers from the 2001 season), and the markets closed sometime between that year and 2005, the actual number of vendors participating during the final season of the market is not known. As is discussed in the following text, one aspect of a market's decline is a loss of vendors.

Thus, some of these markets may have been even smaller just prior to closing. In fact, a recent study demonstrated that a market in distress (at risk of failure) decreases in size and scale (Murray, 2007).

Factor 2: High Need for Products (Lack of Farmer Vendors)

One of the questions in the manager survey asked whether the manager thought his or her market would be improved if it offered more of any of nine product categories commonly sold at farmers' markets. Table 5.8 presents the nine product categories, with the percentage of closed and operating farmers' markets that answered *yes*. These affirmative responses indicate a shortage of products or a lack of diversity in the markets' offerings. All seven markets that closed indicated a significant desire for fruit and vegetables, products considered essential to farmers' markets. More than half the markets indicated a desire for eight of the nine products. In all instances, these markets had a higher need for products than did the markets that did not close. This situation is directly connected to the small size (number of vendors) of the markets noted in factor 1.

TABLE 5.8. Percentage of Oregon markets (closed and operating) responding that more of the designated products would improve the market.

Market Would be Improved by Offering More:	Percent of Closed Markets (n = 7)	Percent of Operating Markets (n = 43)
Fruit	100	56
Vegetables	100	54
Fish	100	72
Meat	100	67
Value-added products	100	61
Cheese	86	81
Bakery products	86	54
Cut flowers	57	35
Nursery products	43	26

Note. The 2002 survey of market managers includes data on seven of nine farmers' markets that were operating at the time of the survey but closed sometime after the survey was conducted.

Factor 3: Administrative Revenue "Have-Nots"

Based on data from the manager survey, the markets that closed were among the administrative revenue "have-nots," as discussed in the Revenue Resources section. Five of the seven markets that participated in the survey collected less than $3,400 per year in administrative revenue (Table 5.9). The range in administrative revenue without temporary grants was $0 to $8,000. With grants included, one market's total administrative revenue was $20,000. This situation, however, was temporary, and the market was forced to close when the grant ran out and the market was no longer sustainable based on vendor fees alone.

Factor 4: Manager Was a Volunteer or Paid a Low Salary

As noted in the discussion regarding market revenue and labor resources, for the majority of markets there is a connection between the amount of money the market collects as administrative revenue and what it is able to provide by way of labor to support the market. Given the situation described as factor 3, it follows that the markets that closed were often those that depended on volunteer managers or that paid the manager a low salary.

Four of the seven managers were volunteers or were paid low wages. Of these four managers, two were volunteers and two were paid between $1,040 and $2,000 per year. One manager associated with a market organization that managed three markets was paid $20,000. One manager was paid $12,000 per year through grant funds. Again, this market was not self-sustaining when grant funds were no longer available.

TABLE 5.9. Administrative revenue of failed markets.

Administrative Revenue	Number of Markets
$1,000 or less	2
$2,600–$3,400*	3
$7,500–$8,000	2

*Categories were created based on actual administrative revenue figures for the seven markets, hence the gap between $3,400 and $7,500.

TABLE 5.10. Manager turnover among failed Oregon markets, 2000–2005.

Existing Markets	2001	2002	2003	2004	2005	Totals
New manager	1	5	4	2	0	12
Returning manager	8	4	1	1	0	14
Total Markets	9	9	5	3	0	–
Percentage of manager turnover*	11%	55%	80%	67%	N/A	Weighted Average 46%

*Rounded to whole numbers.

Factor 5: High Manager Turnover

As indicated in the Market Dynamics section, there is a high turnover rate among market managers (ranging annually from 21% to 44%, with a weighted average of 30%). It is no surprise that this situation is also associated with markets that close. Table 5.10 indicates the number of markets with new or returning managers between 2001 and 2005. All nine markets were operating in 2001, and all had closed by 2005. The percentage of manager turnover for each year ranged from 11% to 80%, for a weighted annual average of 46%—overall, considerably higher rates than the rate for all markets. The nine markets collectively experienced 12 manager changes during the 5-year period; in fact, five of the nine markets changed managers for the majority of the years they operated. Significantly, seven of the nine markets began their final season with a new manager.

ADDITIONAL CONCERNS RELATED TO MARKET FAILURE

Analysis of aggregate data from the survey of market managers revealed some additional areas of concern that could be factors in market failure. These concerns include the lack of market management experience of managers associated with newer markets, a threshold for number of hours worked by volunteer managers, and the erratic management-organization structures of markets in the Small size category.

Years of Managerial Experience and Age of Market

The survey of market managers indicated that the number of years of experience of market managers ranged from 1 to 20. Nearly 37% (36.8) of market managers had only 1 year of experience. Nearly half (47%) had 2 years of experience or fewer. A New Jersey farmers' market study found that 58% of managers had fewer than 2 years' experience, findings similar to those presented here (Govindasamy et al., 1998). This reflects, to some degree, the growth in numbers of markets in recent years. It is also a cause for trepidation. Although a farmers' market could suffer under an experienced but ineffective manager, a study by Oberholtzer and Grow (2003) expressed concern at the correlation between experience levels of the market managers and years of operation for the market: "Thus, in many cases, younger markets—those that could benefit a great deal from market manager experience—lack this amenity for farmers" (p. 24).

There is also a high correlation between managerial experience and the age of markets in Oregon. Newer markets have less-experienced managers, and older markets have more experienced managers ($r(50) = 0.387$, $p < 0.01$). Fifteen of the 16 markets that were 3 years old were managed by managers with 3 years of experience or fewer (Table 5.11). The 10 managers that had 6 or more years of experience were associated with markets that had been operating for 6 years or longer. Seven of the 10 managers with 6 or more years of experience were associated with markets that had been operating for longer than 10 years.

Because of manager turnover between the 2002 manager survey and the closing of individual markets, it is not possible to generate the number of years of market management experience for each manager of the markets that closed. As noted in factor 5, however, seven of the nine markets began their final season with a new manager.

Effort Thresholds for Volunteer Managers

The analysis of market size and management organization structure in chapter 4 revealed that 14 markets (28%) use volunteer managers, while 36 markets (72%) employ paid managers. There is a clear relationship

TABLE 5.11. Years of managerial experience and age of market.

Managerial Years of Experience	Age of Market (Years)				Total
	1–3	4–5	6–10	More Than 10	
1–3	15	2	2	5	24
4–5	1	4	5	4	14
6–10	0	0	3	6	9
More than 10	0	0	0	1	1
Total	16	6	10	16	48

Note. The totals here are for markets, not individual managers. Duplicates are included. There were two "no responses" to one of the questions.

between the size of markets and the status of the manager. Volunteer managers are associated with smaller markets; paid managers are associated with larger markets ($t(48) = -4.917$, $p < 0.0001$). Chapter 4 also revealed that there are no Medium or Large markets managed by volunteers. This is significant.

One implication of this is that volunteers reach a limit in what they are able, or willing, to provide in terms of effort. This issue was examined by analyzing effort for the 14 volunteer managers who participated in the survey in terms of number of hours worked. These volunteer managers average 6.8 hours per week during the market season, ranging from 1 to 15 hours per week (Table 5.12). Ten of the volunteer managers (77%) work 10 hours per week or fewer during the market season. Off-season effort for volunteer managers averages 3.2 hours per week, ranging from 0 to eight hours per week. Nine managers (69%) work 3 hours per week or fewer during the off-season.

The size of markets these volunteers managed ranged from 5 to 29 vendors (including farmers and crafts vendors), with an average size of 14 vendors. Although five volunteer managers (36%) managed markets of 20 vendors or more, nine volunteer managers (64%) managed markets of 16 vendors or fewer. All of these markets fall into either the Micro or Small categories.

TABLE 5.12. Hours worked by volunteer managers during market season and off-season.

	Minimum/ Maximum	Mean	*n*
Hours worked during season	1–15*	6.8*	12
Hours worked during off-season	0–8	3.2	13
Number of vendors	5–29	14	14

*One outlier of 30 hours has been dropped from this analysis because, although acting as a volunteer manager, the manager and the market were associated with a grant supported community development organization.

There is not extensive data documenting what happens as a market grows beyond what a volunteer can manage, relative to the market size and number of hours worked, so it is unclear whether they become overwhelmed or the quality of work declines. These data simply illustrate current practices. Based on these observations, a recommendation for market organizers is to have a plan in place for the transition from a volunteer to a paid managerial position as market size, in numbers of vendors, approaches the midteens and the manager's workload exceeds 7 hours per week during the market season.

Small Market Management Structure

Markets in the Small category (9 to 30 vendors) are in a unique position. Micro markets are sufficiently small that they can be managed in informal ways. But Small markets are of a scale that requires some degree of formality to their management. Medium and Large markets possess an almost predictable set of management practices, but Small markets do not. Small markets are the most unpredictable of the categories in this regard. Small markets mix more formal and less formal management traits. They are developed in some areas, but not in others. Some have additional paid employees, but not a board of directors. They may have a board of directors with officers and committees, but not use some of the standard site-management tools that other markets use. Although their current success could be associated with a specialized adaptation to a community, danger exists if these markets are growing rapidly into Medium sized (or larger) markets. Their greatest risk is not having the management structures in place in order to make the transition to being a larger market. Many may not be structurally ready.

The Small market category is the only category in which there is a significant relationship between the age of the market and the size of the market ($r(20) = 0.498$, $p < 0.05$). The markets in this category grow in size as they age. In general, these markets are growing each year of operation, yet within the Small market size category there is not an increase of management complexity associated with increases in size. These markets seem to be changing, but perhaps in different directions; they mat be growing, stabilizing, or struggling. And remember, these markets often have the least amount of administrative revenue resources available.

Perhaps because of these conditions, Small markets are volatile, with a high percentage closing or changing managers. Twenty markets in the Small size category participated in the manager survey. During the period between when the survey was conducted and 2005, eight of the Small markets changed managers and four closed. A total of 12 of the 20 Small markets (60%) underwent a significant change in leadership or stopped operating. In contrast, no Medium and Large markets closed during the same period, and only eight of these Medium and Large markets (3%) changed managers.

Manager Perspectives on Markets That Fail

Interviews with managers of markets that failed and other managers provide qualitative information that helps flesh out the quantitative data presented in this chapter and by other researchers. The manager accounts note challenges achieving a positive balance between vendors and customers, integrating into communities and gaining community support, and problems within the market organization. They indicate a complexity of issues that swirl within markets and between markets and communities.

Farmers' markets must manage a complex relationship between supply (vendors) and demand (customers). This relationship is different for farmers' markets than for most retail outlets. A viable farmers' market must have enough farmer vendors to attract customers and it must have enough customers to be attractive to farmer vendors. If the market is out of balance it may enter a downward spiral. Burns and Johnson (1996) described this situation:

> Farmers' markets, unlike retail stores, operate both on the supply side, with the farmers, and on the demand side, with the consumers. However, the overall retail marketing dynamic is operative. Consumers wish to have certain preconceptions met when selecting a retail site. If they are not met, the consumers will stop coming. Farmers will go to markets where they are guaranteed selling space and have exposure to enough customers to allow them to sell the majority of their product in an allotted time. When farmer…and customer expectations are not met, both farmers and customers will look for alternative markets (p. 12).

All market managers are concerned with managing vendor and customer expectations to maintain a healthy balance. But many smaller markets seem to be particularly challenged by a general shortage of farm vendors, especially "anchor vendors," those farmers who can provide adequate quantities of a diverse array of products for an entire season. One manager commented on the issue of not having enough farmers and customers at her market:

> Farmer shortage is definitely our most critical issue. And when we do get farmers, often they don't stay for more than a week or two because of the minimal customer numbers. I'd like to consider other market dates, but I'm just not sure that it would solve our problems. I think the demand could be there, but without the farmers and produce to draw in the customers, we have been running in circles for the past four years trying to get the farmers, who'll bring in the customers, who'll bring in the farmers, etc.

Researchers have remarked on the issue of balancing customer and vendor numbers. Burns and Johnson (1996) noted, "managers measure the success of their markets by their ability to attract and retain farmer and customer participation" (p. 14). Oberholtzer and Grow (2003) observed that markets with fewer vendors also had fewer customers. Hughes and Mattson (1992) summarized the relationship between customers, vendors, and labor resources for one market: "More customers result in more vendors, who generate more market fees to pay a coordinator, advertising, and other expenses." They concluded, "This allowed greater input of energy by the coordinator into the market in dealing with vendor issues and market promotion" (p. 8). One Oregon market manager described the challenge of getting the correct proportions of farmers and customers as "magic."

Part of the ability to attract customers and vendors rests with the personnel resources the market has. For smaller markets, these resources are minimal. As one market manager inquired:

> Do markets need to reach a sort of tipping point at which the market is doing well enough to support management that then can work to further stabilize the market? And, if so why do some markets reach that point and go on to success while others never reach that point? Also, why do some markets seem to linger in a 'limbo' state for a few years—not growing very much but managing to survive? What causes a market to either make the jump out of 'limbo' or to finally just shut down?

As markets increase in size, they draw both vendors and customers from a larger geographic area—success breeds more success. Burns and Johnson (1996) observed, "it appears that as the size of the market increases, the market becomes more attractive to farmers from a wider geographic area and the retail (customer) trading area also increases" (p. 16). This has implications for smaller markets. As larger markets draw farmers from a larger area, this process may also draw farmers away from markets they perceive as less profitable. These are often smaller, more rural markets. A California study showed that farmers' market vendor income was associated with community size. Where weekly gross income for rural markets was $291, it was $474 for metropolitan markets (Feenstra & Lewis, 1999). As one market manager recounted:

> I heard from several customers that the prices were high, while I heard from several former vendors that customers weren't willing to spend money at the market. This perception on the part of vendors—that there was little money to be made at the market—combined with the reality of the large number of other markets in the Portland area...made it difficult to attract vendors.

This market manager describes the final season of a western Oregon farmers' market. The market manager participated in the 2002 survey and, at that time, listed the size of the market as 20 vendors.

> The market was located downtown and took place on [a weekday] evening. There were parking problems and tensions with nearby businesses. It was an evening market so it required picking and selling in the heat of the day. Then there was no profit for vendors. Where I would make $120 on Saturday I would only make $30 to $40 [at this market]. The market dropped down to three to four vendors then just stopped.

This account identifies some of the multiple problems this market confronted that led to its eventual closure—lack of parking for customers, tension with local businesses, and choice of market day and time. These problems may have reduced customer numbers, which reduced vendor sales, which in turn reduced vendor numbers. In addition, and importantly, the closing of a Medium or Large size market is rare but we could assume that the number of vendors participating will drop off until the market becomes a smaller size just prior to its failure. This situation is illustrated here with the market dropping from 20 vendors to just three or four vendors prior to closing permanently. It had been a Small size market that became a Micro size market.

Integration into and support by communities is another theme. This account highlights one market's struggle integrating into a rural community dominated by large commodity-oriented farms:

> In our area row crops predominate, orchards are few and gardens are grown only to meet owner needs and to give produce to neighbors. Our region is totally dependent on irrigation. The oppressive heat which lasts from late June to mid-September doesn't help an

> outdoor market. It looks pretty bleak. Next year will be the third and last year to attempt a farmers' market.

This market manager describes a lack of community support as a factor in the eventual closing of a market:

> Early in my involvement with the market, I held a meeting for community members. My goal was to find some volunteers who would help out in various aspects of the market. Despite advertising the meeting only 1 potential volunteer showed up. This proved to be symptomatic of the community's lack of direct support for the market. Although many people professed to appreciating having a local farmers' market, hardly anyone except the existing board members were willing to lend any time or support to the market.

Internal issues within a market may distract market organizers from the business of managing the market. For instance, during focus group discussions one manager commented that issues within the market affect its mood:

> When there is stress between the manager and the board, all the manager's energy goes there. It defines the tone. The customers will know if the governance of the market is distressed.

Lastly, a market manager recounted the inertia among market organizers that contributed to the market closing permanently:

> In the end, I grew frustrated with the lack of support and I could see that the market's problems were not going to be solvable without some fairly substantial changes which the board was unwilling to make.

The quantitative data related to markets that fail describes a downward spiral of interrelated factors of small market size, inadequate product, low revenues, poorly compensated managers and high manager turnover that are seemingly impossible to correct once past a particular point. The qualitative accounts from managers of failed markets describe a complex and nuanced background of issues internal and external to the individual market that ultimately lead to or trigger the downward spiral. The overall analysis suggests a sequence to market decline and failure. Failing markets are small, already at a disadvantage in terms of revenue and personnel resources due to their size. Causes leading to the market's decline appear to be individualized, complex issues that are internal and/or external to the market. These markets enter a downward spiral of declining numbers of customers and vendors, intensifying the problem of inadequate product and leading to even fewer customers. With limits to how much effort volunteers are able work before burning-out, the low revenues to the market interfere with its ability to use paid personnel to stabilize the market. All this occurs in a short period of time; in as little as one year and less than four years for the vast majority of markets that fail.

Areas of Risk

There are two areas of risk associated with characteristics of markets and market managers. Market organizers should be concerned about these areas of risk particularly if they are present in combinations. These areas of risk reiterate the connection between market administrative revenue and the ability to hire paid personnel to manage market operations.

1. **Small markets, markets becoming smaller, and new markets:** Smaller markets are more at risk of failure than larger markets. Since they have fewer vendors, smaller markets are vulnerable to descending quickly into a crisis by a drop in vendor numbers and thus, suffering the repercussions of a lack of products, fewer customers, and low administrative revenue. In addition to having fewer fee-paying vendors, these markets also charge lower stall fees than larger markets and thus, face two challenges to generating administrative revenue. Markets that fail may have been small to begin with or may have been larger but have become smaller through a period of decline. Therefore, any market that is progressively declining in size should be of concern. These markets will also be collecting less money from vendors and will

have a difficult time maintaining management staff. As a separate but related issue, new markets face a higher probability of failure. This is particularly true in their first year and up to their fourth year. Although older markets are not entirely safe from failure, their failure rate is low.

2. **Over-worked managers, under-compensated managers, inexperienced managers:** Manager turnover on an annual basis is high among markets in general at 30% and higher among markets that fail at 46%. Some areas of concern associated with farmers' markets in general include whether and how much managers are paid, thresholds for volunteer manager effort and size of market, new markets and inexperienced managers, and manager turnover. Most managers of markets that failed were volunteers or were paid a low salary. These managers are more commonly associated with smaller markets. In fact, there are no volunteer managers associated with Medium or Large size markets. This likely indicates a limit to the capacity for volunteers to manage markets over a specific size. There is also a correlation between new markets and inexperienced managers, a potentially risky combination.

Chapter 6

It's Harder Than It Looks: The Agro-Social-Economic-Regulatory Ecology of Oregon's Farmers' Markets

So far, the discussion has touched on how farmers' markets are shaped by external influences from the physical environment; large-scale economic, social, and political trends; and variation in access to resources. The discussion has also noted that the manner in which farmers' market management responds to these influences creates the internal conditions that will lead to either stability or stagnation.

Individual farmers' markets originate and operate in the manner they do because of ability of market managers to adapt a market to unique influences in the physical and political environments. This chapter synthesizes findings presented so far, along with additional data; outlines some of the issues faced by market managers; and explores successful adaptations of farmers' markets to physical and political environmental influences.

ECOLOGICAL INFLUENCES AND FARMERS' MARKETS

Political ecological approaches maintain that localized adaptations are influenced by both local and distant physical and political environmental conditions. Applying these principles to farmers' markets offers an effective framework for examining the multiple influences on them, and the resulting outcomes. Farmers' markets are a product of the immediate locale in which they operate and of conditions elsewhere, of which they may or may not be fully aware.

Although farmers' markets are highly influenced by local factors, national and global factors may also be profound influences. *When food appears* in markets, and *what type of food appears*, is dictated in part by factors associated with the agro-ecozone of the region (local environmental advantages and limitations) and the specific climatic conditions for a particular year, or even the particular week of the market season. *How food is handled, measured, weighed, and sold* at a farmers' market reflects state- and national-level regulations. Even the existence of farmers' markets is a response to federal agricultural policy that has favored large farms over small farms, and to international trade agreements that have made wholesale markets unprofitable for small farms. These factors make farmers' markets one of the few viable marketing channels for small farmers.

A number of themes related to a political ecological perspective are apparent in this synthesis. To begin with, farmers' markets operate as *systems*. Systems are defined by Steiner (2002) as "a complex whole [and] a set of connected things or parts" (p. 21). In this way, farmers' markets operate as systems themselves—their parts are connected and affect one another—but they also are a part of larger systems, and so are affected by, and react and adapt to, influences from those systems.

As systems, farmers' markets appear to be stable, but they are undergoing *cumulative change*. As Moran (1990) emphasized, "at any given time, systems appear to be seeking, or be at equilibrium, whereas over time they appear to be undergoing continuous and cumulative change leading to structural transformation" (p. 19). As a result, farmers' markets are experiencing *ongoing adaptation* to their environment, or, borrowing Sheridan's

(1988) description of the political ecology of peasant societies, it can be said that farmers' markets "are neither isolated 'little communities' nor helpless pawns in an international power struggle. On the contrary, they are constantly engaged in a creative dialectic between both local and external forces" (p. xvii). The external forces can be either positive or negative. According to Smith and Reeves (1989), "when analyzing complex socio-natural systems…system processes at higher levels not only set constraints for lower levels but also present opportunities" (p. 77).

A fundamental part of the process of adaptation to physical and political environments is *decision making*. Most individual managers or boards of directors use decision-making procedures in their response to cues from local to global levels of their environment. As Moran (1990) pointed out, "ecosystems approaches have tended to focus on the population and neglected the decision-making activities of individuals" (p. 21). Moreover, as Bennett (1969) adds, with decision making "we refer to coping mechanisms or ways of dealing with people and resources in order to attain goals and solve problems" (p. 11).

A model by Doorman (1991) for analysis of decision making by farmers is applicable to managers of farmers' markets. Within the sphere of decision making, Doorman identified three factors: personal, situational, and external. The information in brackets is added to relate the application of this model to farmers' markets.

> The first, personal, refers to traits inherent in the [manager] as an individual [e.g., education, managerial capacities, tastes and preferences, leadership]. The second factor, situational, refers to the interaction between the [manager] and [her/his] immediate environment [e.g., conditions at the market site, management structure and functions, cash flow, labor]. To a certain extent, these factors can be manipulated by the [manager], and as such create the space that gives meaning to decision-making: the choice between alternatives of action. The third type of factor is external, and refers to those elements that are outside the sphere of influence of individual [managers], and can therefore not be manipulated [e.g., government policy, regional and national socio-political structure]. (pp. 236–237)

Figure 6.1 illustrates that the manner in which a farmers' market appears on a given day is a product of how its manager(s) respond to influences from the greater environment. Some influences are from international and national policy, some from local regulatory issues, some from the specific agro-ecozone of the area, some from the socioeconomic composition of the area, and some from market site conditions.

CHARACTERISTICS OF GOOD FARMERS' MARKETS

In order to examine these environmental influences from the perspective of market managers, one question included in the survey of Oregon farmers' market managers asked: "When thinking about a market like yours, what are the characteristics or traits of a good farmers' market?" This question explored how a manager evaluates another farmers' market. The

FIGURE 6.1. Some environmental influences affecting farmers' markets.

responses indicate what the traits of a well-run market are from a manager's perspective. The question was intended to elicit responses related to the type of market the manager currently operates, in order to avoid clouding the issue by using markets in a different region, of a different size, or with different resources, as their standard. Their responses were analyzed for content, and categories were created based on recurring themes (see Appendix).

Three common themes emerged from the analysis as traits of a good farmers' market: *atmosphere*, *product*, and *community*. The themes are listed as follows (Table 6.1). Numbers in parentheses indicate the percentage of managers that offered a statement related to the theme. For instance, 79% of managers noted *atmosphere* as an important characteristic. Alongside each theme heading are example of quotes from managers relating to the particular theme. For the analysis, each trait a manager stated was recorded, but duplicate traits were counted once.

TABLE 6.1. Traits of a good market based on manager's input.

Examples of Manager Responses	Themes/Traits of Good Markets
• "For customers, an environment that is friendly, inviting, leisurely..." • "Create an atmosphere for lingering." • "Customers just have fun without knowing effort involved." • "Vibrant and vital. Smells, colors, bustling foot traffic."	Atmosphere (79%)
• "Good products. You don't have anything unless you have good produce." • "Variety in product. Quality in product." • "Product mix." • "Variety of in-season produce."	Product (66%)
• "Important to know the community and who you are serving." • "Community ownership. Involve as many people as possible." • "Personal relationships with customers." • "Educated customer base."	Community (63%)

The consensus among managers is that good markets have a pleasant atmosphere, a diversity of high-quality products, and close and multifaceted linkages to their community through a loyal customer base. It follows, then, that markets should give high priority to these traits. Figure 6.2 presents these three characteristics as the core of the farmers' market management ecology. The three traits would seem to be easy to obtain, but the environment in which markets function is fraught with barriers and challenges to achieving these goals.

As illustrated in Figures 6.1 and 6.2, farmers' markets are under influences from a variety of sources. How do they reach their goals when they are surrounded with barriers and challenges? They respond via a

FIGURE 6.2. Traits of good markets: A core of the environment of farmers' markets.

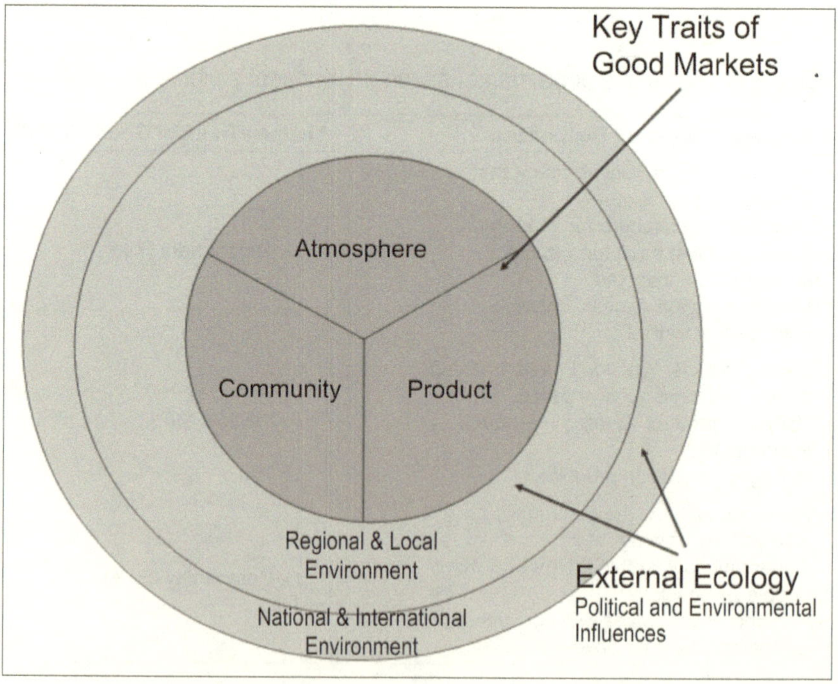

FIGURE 6.3. Management is the interface between environment and key traits.

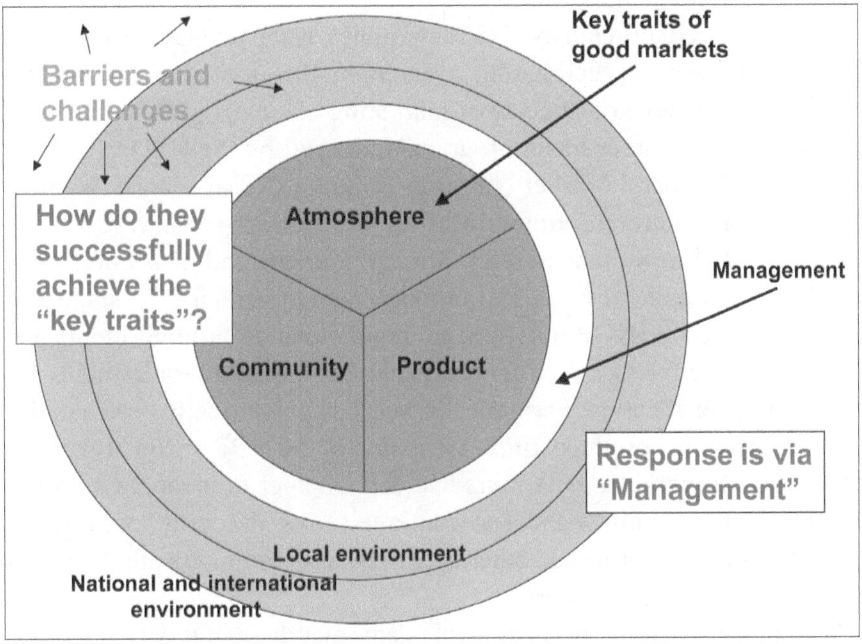

wide variety of measures associated with *market management*, which is at the interface between the external environment and goals. Management and its associated deliberative and decision-making process are the interface between the market's goals and its external influences (Figure 6.3). In this way, management is the act of adapting a farmers' market to local, regional, national, and global physical and political influences (i.e., political ecology).

THE EXTERNAL ENVIRONMENT: GLOBAL, NATIONAL, AND LOCAL

Farmers' markets are influenced by the local natural environment, the demographics of the region, and policy from far away. Not all external influences are negative; some are neutral, and some are positive. What these influences have in common is that they all require a response from farmers' market managers and are largely beyond the control or influence of individual markets.

As discussed previously, USDA policies have favored large farms over small farms, precipitating a situation that forces small farmers to pursue and develop their own marketing channels such as farmers' markets. On the other hand, programs managed by the USDA, such as the Senior Farmers' Market Nutrition Program (SFMNP) and Women, Infants, and Children's Farmers' Market Nutrition Program (WICFMNP) have attracted new customers to farmers' markets and distributed hundreds of thousands of dollars to farmers' market vendors. The National Organic Program (NOP) has specific regulations on farming inputs and cultural methods associated with organic farming and requirements for third-party certification. Farmers' markets have historically been a major venue for organic produce. Implementation of the NOP in 2002 reverberated through markets, as farmers decided whether to incur the expense of certification to allow them to continue to use the term "organic" to attract customers, or to use other approaches to communicate their production system to customers.

On a more regional or local level, state-level food-safety regulations, with enforcement by state- and county-level agencies, ultimately benefit a market by reducing the potential for food-borne illness and the negative economic impact that would accompany an outbreak. Markets are further protected from liability by following government-mandated regulations. On the other hand, many food safety regulations are written for restaurants and supermarkets. Farmers' markets are sometimes faced with the challenge of how to adapt regulations to their temporary sites and still meet standards set for enclosed facilities.

Markets also must adapt to climatic and local demographic and socioeconomic conditions. As noted earlier in this and previous chapters, the types of products available to a market are influenced to some extent by local agro-ecozone conditions. In addition, farm settlement patterns affect whether there are sufficient small farms to supply a particular farmers' market. Local population and socioeconomic conditions affect how large the potential customer base is and whether it consists of retired sawmill workers or investment bankers—one perhaps wishing to buy boxes of fruit for canning, the other wishing to buy a few items for that evening's salad.

These are just a few examples of the types of influences farmers' markets experience from their external environment. All require the attention of the market manager, yet the manager generally has little influence over them. Consistent with Doorman's (1991) decision-making model, most external influences cannot be easily manipulated by a solitary market manager (Figure 6.4).

Market managers are aware that they are influenced by national conditions. The survey of managers asked, "What are some important issues at the national level that impact your market?" Eighty-seven percent of

FIGURE 6.4. Sectors of the environment that influence markets but are difficult for markets to influence.

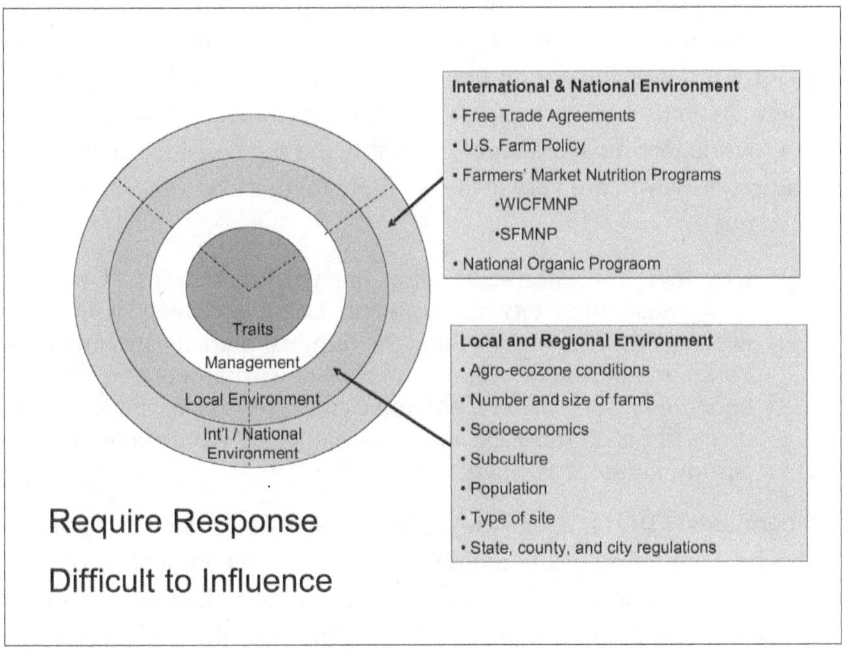

managers were aware of at least one national influence affecting their markets. The most common responses were federal programs: the Senior Farmers' Market Nutrition Program (SFMNP) (74%); the Women, Infants, and Children Farmers' Market Program (WICFMP) (69%); and the National Organic Program (NOP) (18%). Other national influences on their markets included "issues impacting small farms," "free trade," and "biotechnology." In general, managers tend to be aware of national programs (e.g., SFMNP and WICFMP) that have a direct and positive impact on farmers' markets.

Market managers are also aware that they are influenced by state-level conditions, albeit to a lesser degree. The survey also asked, "What are some important issues at the state level that impact your market?" Forty-six percent of managers were aware of at least one state-level issue that affected their markets. Examples of these responses included "preservation of farmland. The style and scale of agriculture is changing and land use planning needs to accommodate smaller parcels" and "could use more consistency in regulations. The fee schedule for nursery licenses is unfair to smaller producers." In reality, state and city regulations have a major impact on markets through food safety, weights and measures, county and city permits, and other issues. These influences are so routine that most managers probably did not consider them. Some managers, however, are very aware of these regulations. As one manager pointed out,

> City level issues have a bigger impact. Frustrations of working with fragmented city government. Lease problems with Parks department who doesn't think the farmers' markets are important. Parking department: Many expenses other markets don't have—hooding parking meters, parking costs, rents to city. Parks department limits number of ready to eat vendors. County requires permit. Parks department requires fee.

Doorman's (1991) decision-making model identifies "situational" factors as being more easily manipulated. This is accomplished by market managers by using management tools—site management, market governance, and labor resources—to influence the three elements of a

FIGURE 6.5. Sectors of the environment that may be influenced by markets.

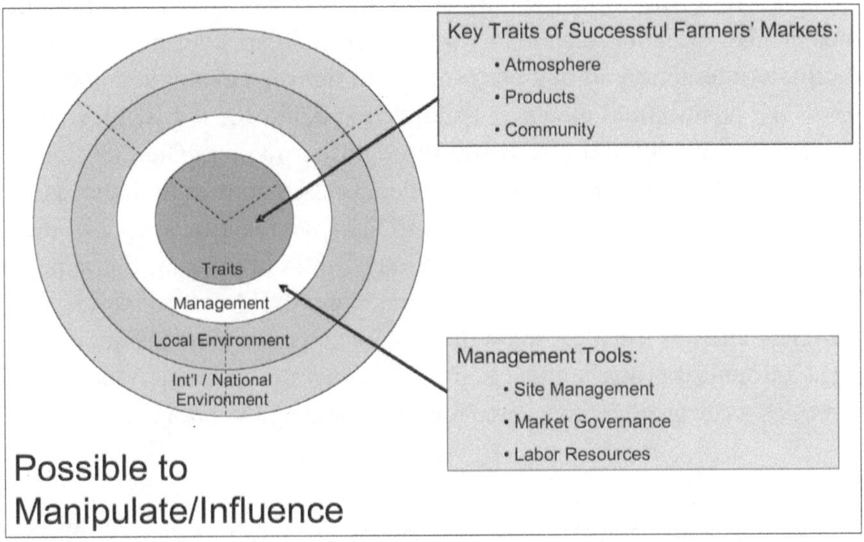

good market (atmosphere, product, and community) (Figure 6.5). What follows is a closer examination of each of the three elements of good markets and the influences on them from physical, social, and policy conditions in local, regional, national, and international environments.

Atmosphere: You Start With a Parking Lot

Here, two aspects of the local environment will illustrate the responses of market managers. As noted in chapter 2, Oregon farmers' markets generally do not have access to permanent sites and have adapted to using often rather stark sites (parking lots and streets) for their markets. Managers identified atmosphere as one of the key traits of a good farmers' market. How do managers respond to the challenge of creating a pleasant shopping and social atmosphere in a parking lot? Figure 6.6 illustrates the conditions that markets face in their local environment: a parking lot for a site, noise, sun, heat, cold, issues relating to access for customers and vendors, among many others. Managers work with the strengths and weaknesses of the market site. They create atmosphere by controlling the form (layout) of the market and where vendors are located. By directing customer traffic to flow in specific patterns, they facilitate interaction among customers and between customers and vendors. By positioning specific vendors in specific locations, they draw customers visually and physically through the market. The addition of live music masks outside noise and encourages customers to linger. Managers also work to influence vendors to improve the appearance of their stalls with attractive signage and displays, thus improving the appearance of the market and adding to the atmosphere. When a market is well designed and has a critical mass of vendors and customers producing the right amount of crowding and activity, it generates palpable vitality and provides a rewarding experience for customers and vendors.

FIGURE 6.6. Management responses to the environment to create atmosphere.

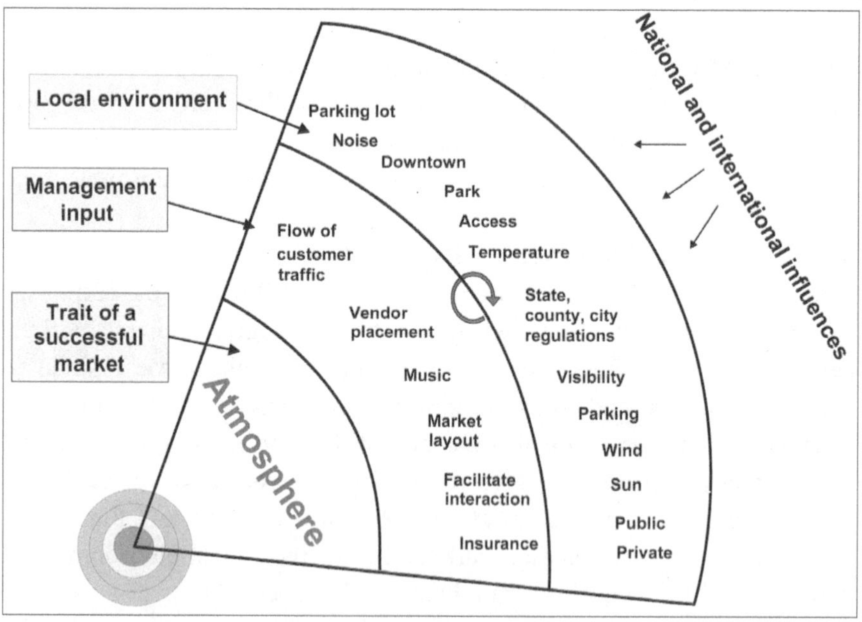

The manager survey data indicate that most markets are well planned and designed to provide opportunities for socializing and sales—markets represent a created atmosphere. The overwhelming majority of market managers (86%) design the layout of their market(s) (Table 6.2). This means that instead of allowing vendors to set up anywhere, the manager gives the market setup a structure: rows, single corridor, circular, horseshoe, and so on. This structure enables customers to flow through the market.

The survey also asked whether managers arranged vendors in the market to meet specific goals. Again, the majority of markets (82%) arrange vendors for specific reasons (Table 6.2). Market managers have several types of goals in mind:

- "Keep agriculture products visible at market entrances. Product mix."
- "Separate products. Corners are important. Arrangement adds spice to market. Traffic flow."
- "Traffic flow. 'Feng Shui.' Customer welcoming—easy for people to come into a horse shoe shape."
- "Look of the market—visually interesting. Separate similar vendors. Keep rows similar. Clump food together."
- "Equal access to each vendor; no dead zones."

Other practical reasons noted by managers for arranging vendors include accommodating individual vendors by separating vendors who do not get along for various reasons, for example; keeping hard-of-hearing

TABLE 6.2. Management input in layout design and vendor arrangement.

Management Input	Number of Markets	Percentage of Markets
Design layout of the market	43	86
Arrange vendors to meet specific goals	41	82

vendors away from the music; separating smokers from nonsmokers; honoring vendor seniority; respecting vendor perceptions of what constitutes a good spot; and so forth.

Comfort is important to the atmosphere of a farmers' market. Chapter 2 discussed the close relationship between open-air farmers' markets and the physical environment. Heat, direct sun, wind, and potentially, rain and cold, are direct environmental conditions at the market site that managers must take into account. Markets often lack adequate shade during the warmer months. Some arrange their schedule to avoid the heat of the day by closing in the early afternoon. As previously noted, one coastal farmers' market is timed to avoid the portion of the day with the highest winds. Another market in a particularly windy area has developed heavy-duty anchoring techniques to keep canopies and signs from blowing away.

Regulations

Local issues can significantly impact markets on a day-to-day basis. Much of what underlies the appearance of a market is an adaptation to state and local regulations. Local health departments enforce state regulations related to food handling and preparation. Samples are an important part of farmers' markets, and local health regulations affect how samples are presented. The Weights and Measures Division of the Oregon Department of Agriculture regulates scales and packaging. For example, in a follow-up to questions posed during a focus session, market managers were asked to list all the regulatory entities they had to deal with in order to operate their markets. Table 6.3 illustrates the regulatory environment of the Portland Farmers' Market through an extensive list generated by the market manager of government and nongovernment entities that it interacts with to operate its markets. The list consists of various state, county, and city regulatory agencies and departments, as well as state and local government and nongovernment entities. Maintaining a positive relationship with all these entities is a time-consuming task that is largely invisible to customers and vendors, as well as to some board members.

TABLE 6.3. Regulatory environment of Portland Farmers' Market.

Oregon Department of Agriculture, Food Safety
Multnomah County Health Department
Oregon Department of Agriculture, Weights and Measures Division
Portland State University: site contract
Portland Parks Department: site contract
Portland State University Transportation Department
Portland State University Facilities Department
Portland State University Grounds Department
Portland State University Commencement Committee
Portland Downtown Community Association
Pearl District Neighborhood Association
City of Portland Street Systems: street closures
Portland Fire Department: propane permit
City of Portland: noise permits
Ecotrust: private contract for Thursday market site
Department of Transportation: meter hoods, lane closures, parking
Police, ADA, Oregon Liquor Control Commission: approval and permits
Rose Festival Association: uses the market's regular Saturday space for its art show; the market requires its permission to set up in the next block
The Multnomah Food Policy Council: works with policy that affects market operation
Portland Development Commission: currently works with the market to build a relationship for the future
City of Portland: sign permits

Product: Farms and Food

Managers identified *product* as another of the key characteristics of a good farmers' market. In simple terms, product is what is sold in the market. Food items are the predominant product type sold and will be the focus of this discussion. Managers identified product quality and product mix as important indicators for this segment. One of the perceived differences between farmers' markets and retail food outlets is a reputation for local, fresh, high-quality produce. Managers seek out vendors who can provide this. "Product mix" refers to the array of types of products. A goal would be to offer a full spectrum of vegetables and fruit along with other products such as cut flowers, nursery plants, bakery goods, meats, fish, and cheeses. Figure 6.7 illustrates the types of conditions that affect quantity, diversity, and quality of products, and the types of management responses used.

FIGURE 6.7. Management responses to the environment to obtain product.

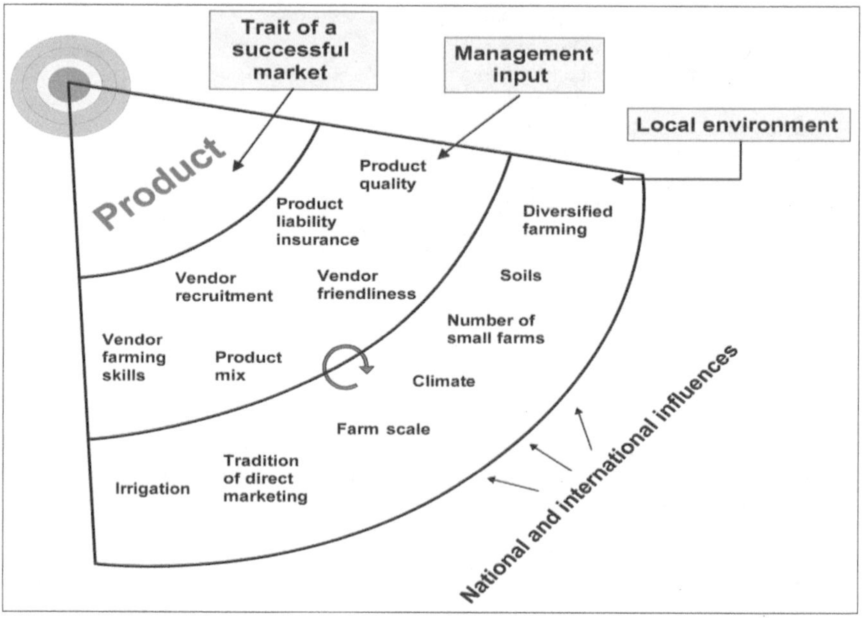

The agro-ecozone of a region largely determines what types of crops can be grown there. Soil types, the number of frost-free days, amounts of rainfall and when it occurs, and daytime and nighttime temperatures during the growing season are among the many factors that influence the array of crops that can be grown in a region. Because of regional growing conditions, no citrus or avocados are grown in Oregon, but crops such as caneberries and apples flourish. There are marked differences in growing conditions between subregions of Oregon as well. These conditions can determine how a market defines "local."

One factor influencing the supply and types of products available to farmers' markets is the scale of farms in the area. Some agricultural areas are locked into producing large quantities of commodity crops (wheat, potatoes, onions, etc.). The farm-settlement pattern in these areas tends to favor large farms specializing in a narrow range of crops. In other regions, often in addition to larger-scale farms, there is a critical mass of small farms whose method of farming often includes a more diversified approach; a greater focus on producing smaller quantities of a wide variety of crops, including unusual or specialty crops; and using nonwholesale market channels (roadside stands, U-pick, farmers' markets, community-supported agriculture, etc.) in order to be profitable. The farm settlement pattern in an area is the result of political and environmental influences on farm production and marketing. It is important to note that the number of small farms is not the key; it is rather the number of small farms that have the physical resources (soil, water, and climate) to engage in farming activity and whose operators are engaged in the type of farming and marketing system that provides products for farmers' markets and other direct-marketing channels.

Farmers' market managers work within the limitations of what can and cannot be grown in their areas and, sometimes, of the limited number of farms to supply produce. Managers influence products offered in their markets by recruiting vendors to add a specific product or to increase the quality of a specific product. As previously noted, they also try to increase vendor sales by influencing them to improve their signage and displays; to smile and be friendly to customers; and to extend vendor growing seasons or increase product quality by disseminating information on farming techniques they have heard or read about, such as using hoop houses to increase the growing season in small fruit and vegetable production systems.

Often, managers will accept most farm vendors in order to increase the quantity of product and perhaps diversify the prices offered in their markets. In some instances, when a market has attained its size goal, a manager will substitute some vendors with others who produce a higher quality product or who offer new products that have high customer demand. Ultimately, these managers are faced with sustaining the market by serving its customer base. As one manager whose market had attained its goal for size commented, "The dynamics change when the market is full. I can concentrate on quality and diversity [of products]. It is better for the health of the market to focus on quality and diversity rather than the seniority of vendors."

Table 6.4 shows the basic array of products offered at farmers' markets. Almost all markets offer products for which farmers' markets are most noted: fruit, vegetables, and cut flowers. In addition, nursery products such as bedding plants are offered at all markets. Bakery products and processed food products (jams, salsa, and dressings) are also very common. More recently, products such as meat, fish, and cheese are emerging.

However, when managers were asked whether they thought their markets would be improved by offering more of the same items, a majority said "yes" for all products except nursery products and cut flowers, indicating a continuing high demand (Table 6.5). This is true even for the products seen as the foundation of farmers' markets—fruits and vegetables. Managers feel that supply is not meeting demand for these products, or they feel there is a need to expand and diversify the variety of these products in their markets.

Notably, the demand for products is influenced by where a market is located in the state. Table 6.6 breaks down demand for individual products in Oregon by region. Here, "high demand" is arbitrarily defined as having 50% or more of the markets respond that offering more of the product would improve their markets. It is striking to note that two regions of the state have a very high demand for most of the typical products offered at farmers' markets. Of the nine products listed, markets

TABLE 6.4. Products market currently offers.

Food or Farm Product	Number of Markets	Percent of Markets
Fruit	50	100
Vegetables	50	100
Nursery products	50	100
Cut flowers	48	96
Bakery products	44	88
Processed food products	41	82
Meat	22	44
Fish	20	40
Cheese	16	32

TABLE 6.5. Market would be improved by offering more of the following products.

Farm or Food Product	Number of Markets	Percent of Markets
Cheese	39	78
Fish	37	74
Meat	35	70
Processed food products	31	62
Fruit	29	58
Vegetables	27	54
Bakery products	26	52
Cut flowers	16	32
Nursery products	13	26

TABLE 6.6. Demand for products and percentage of markets by region.

Farm or Food Product	Eastern Oregon	Coastal Region	Southern Oregon	Willamette Valley	Portland Metro
Fruit	100%	83%	75%	46%	41%
Vegetables	100%	83%	75%	46%	35%
Nursery products	100%	33%	13%	26%	12%
Cut flowers	100%	50%	50%	26%	12%
Meat	75%	16%	38%	80%	82%
Fish	100%	67%	38%	86%	76%
Cheese	100%	50%	75%	80%	88%
Bakery products	75%	83%	38%	46%	53%
Value-added food products	100%	75%	75%	72%	47%
Demand Ratio	9 of 9	7 of 9	5 of 9	4 of 9	4 of 9

in the eastern portion of Oregon, east of the Cascade mountain range, have a high demand for all nine. For the coastal region, seven of the nine products have a high demand. Southern Oregon has a high demand for five of the nine products. The Willamette Valley and Portland Metro regions have a high demand for four of the nine products. Within these regions there is an interaction that occurs between agro-ecozone conditions and farm settlement patterns that produces the characteristic types of products for farmers' markets in areas of higher population density. For the eastern Oregon areas, the agricultural production regions have the capacity for growing a wide variety of crops, but they tend to be oriented toward larger farms producing a few specific crops directed at wholesale commodity markets. These areas also have lower population densities. Agriculture in the coastal region is dominated by dairy and other livestock production and has limited fruit and vegetable production. As one coastal farmers' market manager pointed out, "it is hard to find a flat spot." This area also has a relatively low population density.

In contrast, the southern Oregon region is dominated by the Rogue River Valley, and the Willamette and Metro regions are dominated by the Willamette River Valley. Both of these valleys are notable agricultural production areas and have farm settlement patterns that include a high number of small farms located on good soils with access to irrigation. These areas also have higher population densities than the coastal or eastern Oregon regions. The bottom line is that they have the population base to create a high level of demand; they are located in areas with fewer limitations on agricultural production; and, perhaps because of the high population density, there are a sufficient number of smaller farms located on good agricultural land to supply products for farmers' markets. Consequently, farmers from the Willamette Valley regularly attend markets on the coast and in eastern Oregon. In short, a characteristic that is considered key to being a good market—a mix of high-quality products—is influenced by where the market is located, which is, in turn, influenced by the physical environment, farm settlement pattern, and population density of each region.

Community: More Than Customers

Along with agro-ecozone variations and local adaptations to sites and regulations, markets exist in communities that consist of different population sizes, a diversity of subculture groups, and varying socioeconomic conditions. Within their communities, markets strive to foster a dedicated customer base and garner the support of decision makers and associated business owners. Figure 6.8 illustrates the *community* segment of the model. Community aspects of farmers' markets include relationships with customers, relationships with community groups and decision makers, and "being the community" by reflecting the community.

FIGURE 6.8. Management responses to the environment to build community.

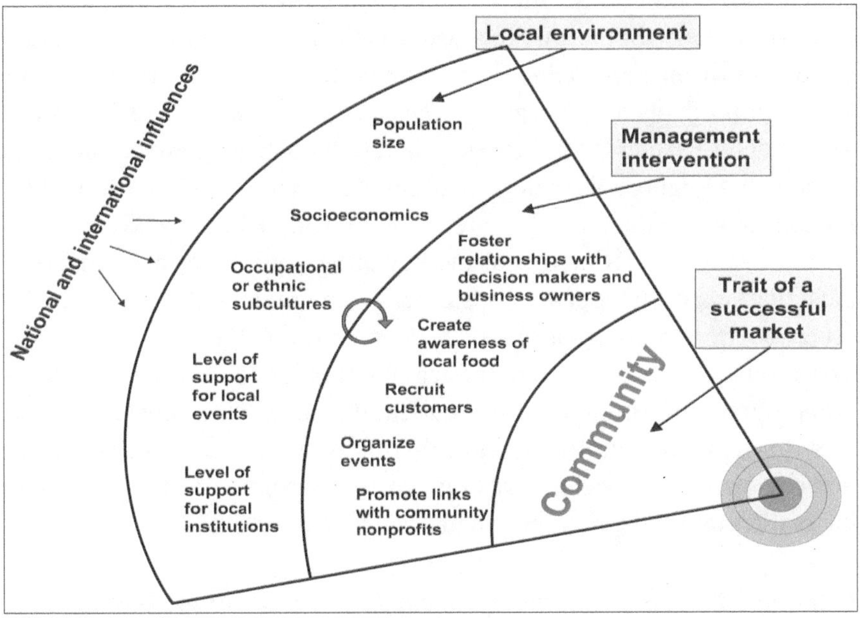

Relationships With Customers

Farmers' markets are highly social entities. As noted previously, managers attempt to create an atmosphere that encourages lingering and socializing. It is common to see people engaged in extended conversations in farmers' markets. All markets encourage family members to attend, and many even cater to children by offering special activities. Chapter 3 noted that a major factor in the growth in numbers of farmers' markets during the 1990s was community building. In this way, farmers' markets are used as vehicles to revitalize and enhance the quality of life in cities and towns by attracting large numbers of people into a specific area for social and economic interaction.

Recruiting customers by generating awareness of the market is a constant concern of market managers. They look for ways to efficiently advertise their markets. For instance, some farmers' markets advertise in their local newspapers during the market season, promoting the current "hot" product in the market, such as berries, tomatoes, sweet corn, or melons. Some markets use e-mail distribution lists to remind customers of an approaching market day and advertise special events and what products are in season. Many farmers' markets use Web sites. The Oregon Farmers' Market Association publishes an annual guide to markets in Oregon. It is widely distributed to a variety of outlets, including tourist venues and local reference and resource organizations.

Because local food is one of the key attractions of farmers' markets, most markets try to appeal to customers who are interested in the freshest products, as well as those who feel a commitment to supporting local agriculture. For example, one farmers' market launched a public education program informing customers about the power they have to support local farms with their purchases. Other promotions, such as demonstrations by chefs and special days (e.g., berry and tomato festivals), extend the market into the community and draw customers.

Relationships With Community Groups and Decision Makers

Farmers' markets foster close relationships with other community groups. Eighty percent of Oregon markets (40 markets) allow booths for community groups. The inclusion of community groups potentially decreases the revenue for markets, but it increases the support for the market by groups that use it as a venue for interacting with the public; it increases the value of the market as a center of community activity and bonds the market with the community. Community groups see the market as an important setting for exposing their organizations to a large sector of the public.

Relationships with local decision makers and the business community are key to maintaining a market's standing in the community. Demonstrating the value of the market as a community gathering place and as an engine of economic development for both local farmers and the associated business district is an important role for managers. During focus-group discussions, one manager commented that he considers "local decision-maker support as one of the most difficult things for a new manager to impact. Many managers don't have the time." Another manager noted, "farmers have difficulty integrating the 'community' element into their lives. If you don't manage this third, hardest piece, you might lose the site."

Being the Community

The communities in which markets operate are a diverse blend of subcultures, socioeconomic conditions, and population levels. Each community is unique. During the focus-group discussions, one manager observed that "a successful farmers' market responds to the community and what the community wants." Another manager added that markets reflect the "personality" of their communities.

By blending the resources at hand (sites, vendors, and customers), markets resemble the communities in which they operate. For example, one rural farmers' market operates in a community with a small population. The market's goal is to provide income for its vendors and to bring people in the community together. The site is on a lot near a local café. Most of the 12 vendors are on a first-name basis, and the market is not a stickler for rules. The manager is a volunteer who receives a free market stall in exchange for her efforts. The market serves retired and low-income families. Food products are surplus from gardens and part-time farms that grow produce as a sideline. Just about every local product is welcome, including clothes, jewelry, and, the manager admits, "too many cheap crafts," but she goes on to say that she "can't resist allowing them in." The pace is comfortable and friendly, and the conversations are long. This market has been specifically designed to celebrate its rurality.

A market in Portland, on the other hand, occupies one of the more unique sites in Oregon. Situated in the downtown area, the site is small and very urban. The market attracts vendors from a number of the area's best farms, some coming from nearly 100 miles away. The site barely holds the 45 vendors who suffer the inconvenience of difficult unloading and are then crammed together more closely than they might tolerate at other markets. However, the market is very profitable for farmers. It opens at 10 a.m., and the early customers are typically residents of the downtown area who stock up on fresh products they would otherwise have trouble accessing in downtown grocery stores. Just before noon, the pace quickens as workers on their lunch breaks arrive to grab something to eat or to shop for dinner. Chefs from nearby restaurants demonstrate how to prepare meals with items available from the market that day. The small space becomes crowded. Money, vegetables, fruit, cheese, and bread are exchanged quickly. People squeeze by each other, bags of produce held over their heads. The commotion builds. Customers who have completed their shopping flow out of the entrances, laden with bags, disheveled but energized. By 1:30 p.m. the crowd has returned to work, and the market staggers toward its 2 p.m. closing time. These rural and urban farmers' markets are very different, yet both reflect the communities in which they operate.

INTEGRATION OF THE MARKET MODEL

Figure 6.9 presents a visual model of the management ecology of farmers' markets. As with any visual model, it is a simplified representation. Conditions that influence farmers' markets from the global to local levels (the external environment) form the outer portion of the model. The key characteristics of a good market are defined in the large inner ring, with the management ring forming the interface between the key characteristics and the rings representing external environment. Lines between segments are dashed to signify that they are permeable. The model represents farmers' markets in an environment made up of influences from natural and political conditions at the international, national, and local levels. Through management, markets adapt to these conditions. Their adaptations are visible: they create an atmosphere conducive to socializing and sales; they procure a variety of high-quality products; and they build community support via a loyal customer base and integration into local social and economic spheres.

In reality, there are overlaps and linkages between the *atmosphere*, *product*, and *community* segments. For example, during the focus-group

FIGURE 6.9. Ecology of farmers' markets.

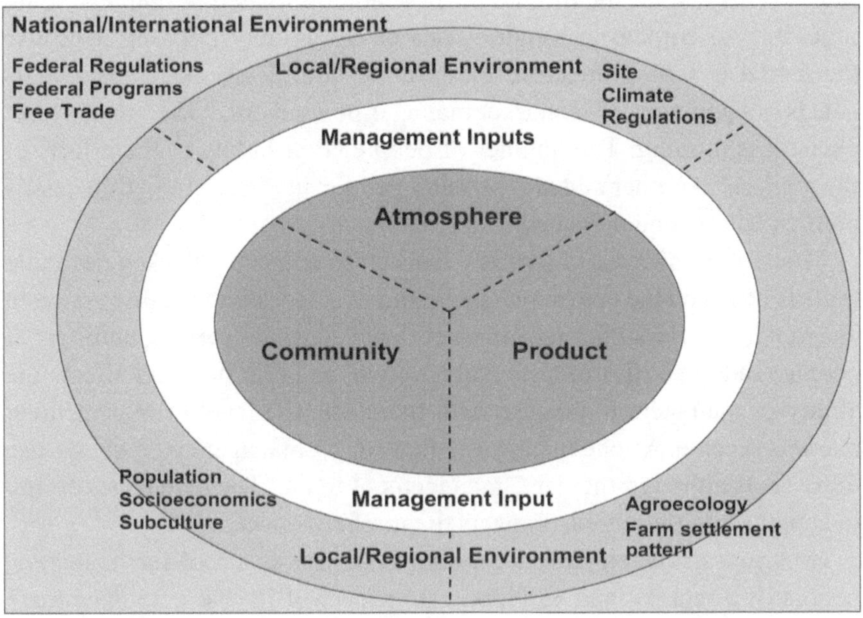

sessions, one manager responded to a draft of the model noting that she sees vendor placement related more to *product* than to *atmosphere* because it is related to sales: "Sales-wise it [the market] looks really inviting; it tells customers 'I'm here to sell.'" Another manager added, "There is not a clear, defined line between *atmosphere* and *product*. When you set the market up, agricultural products are most visible but you want more intriguing products to draw customers in. For nonagricultural products, you need to think what kind they are and what kind of customers they pull in. That promotes customer flow and enhances customer numbers." Another example of how these sectors overlap is that there are often perishable products left over on market day that are sent to local food banks, which demonstrates a link between the *community* segment and the *product* segment. Some markets may emphasize one of the segments more than others. For example, one market manager commented, "Markets have different missions. For me, it's all about sales." This echoes the discussion in chapter 3 related to markets being organized to emphasize vendor sales or community building. Because the model is a simplification, some of the complexity of farmers' markets is not apparent. One market manager pointed out, "The human asset element is missing. The number of people contributing to the efforts of the market." Another added, "Missing vendor success. Market success is predicated on vendor success."

Much of the success of farmers' markets in achieving the key desirable traits is based on the use of management tools. Recalling the discussion in chapter 5, individual farmers' markets have access to varying numbers of people, and quantities of time and revenue, and this situation affects the ability of managers to manage and, therefore, the level of management markets receive. As one manager indicated, "A manager can't always use these tools although they are 'management tools.' The manager's ability to actually use these tools impacts the market's success."

One goal of developing the model is its use as a tool for managers, boards of directors, and vendors. One aspect of its use is to encourage those associated with farmers' markets to see the "big picture." One manager said, "This shows all the things that the people like and what happens behind the scenes to make all that happen." Another said, "The three elements are a three-legged stool: some make others possible." Other managers recognized the use of the model as a planning or educational tool. Regarding focusing management on the traits of successful markets that market managers identified, one manager said, "I think this is a useful tool. I see the center of the center; you get more leveraging the center of the circle. Put your effort where you can have an impact." Another noted an appreciation for the work she does: "I feel encouraged by seeing what I do have an influence over. And, I want it produced as a poster for my board [of directors]!" Lastly, this manager summarized the potential value of the model: "This information could have a lot of meaning to farmers' market boards and opportunities for educating boards and guiding market development or market resources."

November Vignette

Saturday, November 9

7:56 a.m.

It's nearly the last Saturday farmers' market of the season. When I arrive at the market site it's raining hard. The market opens in slightly over an hour, but only a few vendors are here. During August, there might be 40 or more vendors here by this time. Today, there are only 6 vendors setting up: 3 of the area's larger organic farms, a fruit farm, a bakery, and a fish seller. John with the "Nut House" arrives about 8:05 and parks in a space with a remnant of cracked filbert shells on the pavement, indicating that he parked there last Saturday. Sandy, a baker, says she got up at 6:00 a.m. The rain picks up. She says, "I should have quit November 1st!" Another vendor: "Well, this defies reason, doesn't it!"

8:15 a.m.

There are nine vendors setting up. Jokes abound regarding the weather, the hour of day, and so on. A large puddle of water cascades off a canopy, briefly distracting the vendor setting up beneath it.

8:30 a.m.

There are 11 vendors. The manager says vendors have been arriving later the past several weeks. The rain slows arrivals even further. There will be a lot of "no shows" today…

Noreen arrives in a late '60s van. She scans the market site—the manager sees her and calls out. They stand under another vendor's canopy, pointing at various potential stalls. The vendor pulls into a spot. She and her husband begin the process of setting up. They sell apples. Their stall is placed away from Rita, the other apple vendor.

8:36 a.m.

One of the large organic producers is still setting up. Larger produce vendors arrive early because of the amount of time it takes to set up. The process involves setting up a large canopy first. Then, stacks of produce boxes are moved in as display tables are set up. Colorful table covers are added, then produce is arranged. Stacks of empty baskets await placement and filling.

Noreen's husband walks slowly toward a car parked on street, glancing at Rita who is setting up her apple stall. Noreen sets up their stall using the back of the van, a standard canopy in blue, and two card tables placed to avoid dripping from the canopy.

8:38 a.m.

The first early customers arrive. Noreen stacks boxes of apples next to the tables, places colorful covers on the tables, and places a scale in the center of one table. A folding chair leans against the van.

8:45 a.m.

"Morton River Farm" arrives. "Gates Farm" employees say they got up at 6:00. They say they don't have to get up as early as they do in summer.

They arrived here about 7:00. It takes them two hours to set up. Their display is elaborate.

A man strolls with a single leek in a plastic bag; his wallet out, he peers at Noreen's apples. He stops and talks to "Morton River." Walks on. "Morton River" has a simple setup: a single table with a parasol, a dozen egg cartons in a basket, an ice chest full of broilers. A farmer is sorting winter squash and puts her farm name on the small display. Bill, of "Morton River," will sell the squash for her.

9:00 a.m.

It is opening time for the market and most vendors are still setting up—methodical, not frantic. Customers select kale and potatoes at "Gates Farm" amidst unpacked boxes. An employee weighs carrots while another employee continues to move boxes.

The farmer consigning squash with "Morton River" puts prices on the box with dayglow pink paper. Noreen has boxes of apples and bags of apples displayed. She chats with a customer. Money is unfolded as bags of apples are weighed. The folding chair still leans against the van.Customers depart with bags of produce and apples and walk briskly to cars. It has stopped raining. A man and a woman with a plastic tote approach "Morton River." Bill consults a ledger and palms his calculator. He places one, two—selecting—three broilers, then a fourth broiler into the tote. He calculates… The farmer consigning the squash stands by. Bill scans the horizon as he speaks. The farmer hands him a paper, he glances at it, and the farmer departs. The customer is still writing a check. A woman with beets and parsnips navigates the parking lot landscaping to get to her van.

9:13 a.m.

Bill strolls, smiling toward Noreen. They exchange comments while Bill paces, facing his stall. An employee at "Great Roots" sits on a produce tote and removes his yellow bib rain pants.

A balding man in a rain parka and shorts walks away from "Great Roots" with a box full of produce—a CSA pick up.

A "Gates Farm" employee hangs produce bags from their canopy—nearing end of setup. Bill opens a Pepsi, and strolls back to Noreen.

A woman departs with a huge bag of carrots. She strains toward her car. The bag must be 25 pounds at least.

9:15 a.m.

The manager strolls slowly through the market. She stands with a clipboard under her arm. Her hood is down now. Thirteen vendors are spread around the lot in a summertime layout, huge spaces between them. It is starting to rain again.

9:30 a.m.

Rita's employee. in a green rain suit, offers samples of apples. As I chat with Noreen, a man on a bike rides up, selects a large, yellow and red apple, and places it on the scale. Noreen says, "Now that's a Jonagold..." and explains its tartness and sweetness.

As the rain falls, the customers pick up their pace between vendors. Customers walk diagonally into the market, cutting through areas that would have had vendors occupying them two months ago.

"Great Roots" is offering red potatoes, leeks, russet potatoes, various bell peppers, small heads of cabbage, beets, turnips, yellow onions, butternut squash, delicata squash, a few plum tomatoes, carrots, pumpkins, ornamental dried corn, lettuce, and greens.

9:36 a.m.

The rain lets up, then picks up again. A few umbrellas pop open. There are about 25 people shopping. Business is slow but steady.

A "Gates Farm" employee pushes pools of water from the canopy. Behind a vertical display of vegetables layered in color and size is a hodgepodge of boxes: some standard plastic totes with covers, other boxes showing use; *Ladybug* brand, plain brown, and plain white; thin and thick; with and without lids.

Bill talks with a customer as a small boy with his hood up and his hands pulled inside the arms of his coat stands stiffly by.

A large woman with a raincoat like a tent pushes an empty, foldable shopping cart. A woman tastes an apple and watches as another customer talks with Noreen, waving her hands. She walks away.

9:53 a.m.

People appear with a baby in a stroller and a baby in a harness. Bill has fingerless wool gloves. He reaches into the back of his station wagon, grabs a dozen eggs, and opens the lid. Two women customers lean forward to see the eggs. The rain has stopped.

You can hear the crack of a hand-cranked nutcracker…traffic on a nearby highway…car doors closing…whistling.

A "Great Roots" employee samples Noreen's apples. They talk. A man conversing with a "Great Roots" employee looks quickly in the direction of the sound of water hitting the pavement. The water is draining from a pool on the canopy. Another "Great Roots" employee fills a bag with russets, weighs them, and trades them with Noreen for apples.

In the market are four large organic produce vendors; four apple growers; two filbert growers; one smoked salmon vendor; one "mixed" vendor selling baked goods and vegetables; and one vendor selling eggs and broilers, along with the consigned winter squash.

10:03 a.m.

Noreen reaches for a hat on the dashboard of the van. An oyster vendor shows up and unloads a large, white, pop-up canopy from his pickup. He makes a long, stretching, yawning noise upon arrival.

Hands in pockets, shoulders stiff, another man walks with two boxes of apples, arms fully extended under the weight. Two kids follow him.

The oyster canopy is partially up. The seller gets plastic bags from his truck and hangs them from a high point in the canopy, then extends the canopy legs.

Shoppers carry coffee in insulated cups. Women stroll with empty canvas shopping bags.

The oyster vendor opens his tailgate and moves long, clean, white, Styrofoam boxes to the back. Some are marked "ice," some "blue." One is marked "Haye's Grill Tualatin" in grease pen.

A woman in a toque and rain parka chews on a large carrot. She has more carrots in a plastic bag. Her friend stops at Noreen's. Kids play in the parking lot landscaping.

A large delivery truck for a window manufacturer was parked in the lot this morning, forcing "Spring Mountain Farm" to move from its usual location. This would have been a big problem in the summer.

"Spring Mountain Farm" has a selection of lettuce, fennel, cauliflower, tomatoes, and standard winter vegetables. "Donson Farm" has persimmons, standard winter vegetables, potatoes, carrots, greens, and fennel.

10:25 a.m.

Twenty shoppers are in the market. A woman walks deliberately, budgeting the distance and the strength of her fingers to carry four gallons of apple juice. "Donson Farm" employees tell me that they woke up at 5:30 and arrived at 6:30. They laughed when they told me this.

10:35 a.m.

There are 25 people shopping. An intense man wearing two short-sleeved shirts and a fleece vest quickly arranges newly purchased apples in his shopping bag.

A man nearly out of earshot notices a "Gates Farm" employee calling him. He quickly turns as the employee walks toward him, smiling, spinning a plastic bag of salad mix to close it.

10:49 a.m.

At Noreen's stall, a woman with four canvas bags laughs loudly but politely. A "Great Roots" employee adjusts turnips. The oyster vendor has visited Sandy's stall to buy a cookie. The rain has stopped and it is cloudy but bright. The oyster vendor talks on a cell phone.

A gust of wind lifts the oyster canopy, and as a "Gates Farm" employee goes to assist, two legs fall from the "Gates Farm" canopy. The "Gates Farm" employee throws a rope over the frame of the canopy and appears to be preparing to rig some sort of weight to stabilize the canopy, but is distracted by the need to help a customer. Later, two employees look at the rope to see if it can be attached somewhere.

Bill carefully writes an order for broilers into his book. The wind gusts up again. The market manager strolls by, eating filberts. Demands on her time are low. The "Fine Fruits" vendor moves apples—two per hand, jammed between her fingers—from the van to the display.

11:12 a.m.

Twenty-two shoppers are in the market. A couple smiles and says, "No, thanks" to Noreen's offer of a sample. Then they reconsider, and Noreen lifts the cutting board toward them. Another woman samples, too. They chew, comment, and nod affirmatively.

Ladybug brand boxes read "Oregon's Own Organically Grown." Other boxes read "Northwest Fresh Fruits," "We grow everything we sell," and "Free range eggs."

A man barely breaks stride to turn toward Noreen while refusing a sample. A man with him stops and accepts a sample as his friend keeps walking; he chews thoughtfully and nods positively. A couple walks into the market briskly, scanning, squinting to see what vendors are here.

11:23 a.m.

There are 24 shoppers now.

A woman with eggs hugs a "Gates Farm" employee. A man with a baby carrier clutches a dozen eggs and carries a persimmon delicately between his fingers; his arms are full. Bill still has two spaghetti squash and six acorn squash in wooden trays on the pavement next to his table. Eggs and broilers are the main feature. He loads birds into plastic shopping bags. A customer makes two trips—three bags each time, six birds in each bag.

11:57 a.m.
It's raining again. There are 13 shoppers in the market. A light gust blows a cascade of pooled rain off the "Gates Farm" canopy. The market manager collects fees. Water on the "Great Roots" canopy falls on a customer looking at russet potatoes. Noreen wipes up rainwater that has blown in on her tables.

12:00 p.m.
Bill tries to leave his stall to buy something to eat, but a customer catches him as soon as he leaves. Another customer arrives and stands patiently under Bill's parasol. Bill returns. Later, Bill writes something on a piece of note paper and tapes it to his table so customers can see it. He opens the brownie he bought. There are no more eggs. It starts raining hard.

12:30 p.m.
John, with "Nut House," has left.

12:45 p.m.
Noreen starts packing up. Bill calls out, "Hey, you're leaving early, you can't do that. The customers will be flocking here any minute." Another apple grower starts packing up, and, at 12:47, the smoked salmon vendor does too.

12:48 p.m.
There are four shoppers now. Rita has packed up. The oyster vendor is in a full rubber rain suit. Bill says he would leave but he's consigning something for another farmer and she won't show up until 1:00.

12:51 p.m.
"Gates Farm" employees begin sorting boxes and packing. The oyster vendor backs his truck around for easier loading. There are six shoppers in the market.

12:54 p.m.
Bill starts sorting and putting away boxes, watching for the other farmer.

12:55 p.m.
"Great Roots" still has a customer. "Fine Fruits" is beginning to tear down its display. "Gates Farm" moves its truck into the market.

12:57 p.m.
"Great Roots" is removing price signs from baskets. Someone tries to buy something at "Gates Farm."

1:00 p.m.
One customer is still shopping amidst rain and vendors packing. She's able to buy oysters.

Everyone is packing up now, orderly and measuredly. Bill settles with his consignee.

1:12 p.m.
Sandy walks back to her stall from "Donson Farm" with bundles of greens.

Chapter 7

Concluding Observations and Recommendations

This book examined conditions associated with the success and failure of individual farmers' markets. The research findings are intended to provide information for market organizers and practitioners who work with them, to assist with their decision making and strategic planning. I am hopeful that these findings, coupled with the recommendations provided in this chapter, will contribute to improving the viability of farmers' markets.

A Broad Discussion of This Research

This research was conducted during a period of seemingly boundless interest in and support for farmers' markets. Because there is a high level of support now—and knowing that support can decline—it is important to use this opportunity to put farmers' markets in a strong position for future success. This can be accomplished by focusing on two main areas: internal management issues and external public policy issues. This includes enhancing the efficiency of management systems for farmers' markets *and* increasing their awareness and ability to manage public policy to favor farmers' markets—or at least to prevent it impacting upon them. Although there is a lot more to be done, this book has added several contributions to our knowledge of farmers' markets:

1. It has identified "community building" as one factor driving growth in numbers of farmers' markets.
2. It has examined the management organizational structures used by farmers' market organizers and associated sets of these structures with market size categories.
3. It has revealed that high numbers of markets fail and identified some of the causes for failure.
4. It has created a visual model of management ecology that illuminates external influences on farmers' markets and how successful markets adapt to these influences, thus enhancing the decision-making and strategic-planning capabilities of market organizers.

These contributions will be elaborated on in this section and other sections of this chapter.

Intriguing Themes

Some themes emerged during the research that are intriguing and warrant further research. The first theme deals with the complex relationship between the supply of and demand for farmers to participate in farmers' markets, coupled with an evolution of markets and market-customer tastes. The second theme deals with the repeated theme of "community" in relation to farmers' markets.

Farmers and Farmers' Markets

A swirl of issues emerged surrounding the relationship between farmers and farmers' markets. These issues involve the supply and demand of both farmers and farmers' markets, and point to the complexity of the relationship of farmers and farmers' markets as individuals and collectives. Three potentially interrelated issues illustrate this:

- Some very successful urban markets have stated that they have reached their goal for size (in number of vendors) and are now concentrating on expanding the product mix and improving the quality of products offered. This means that these markets can become very discriminating in their choice of vendors. In some instances, farmers may be denied participation in a market in which they had previously participated. This level of market success reveals a tension between the needs of farmers for a marketing channel and the needs of the market to sustain itself by growing its customer base.
- Conversely, as the number of farmers' markets increases, there is a potential shortage of farmers to fill all of them. Managers of many markets (particularly those in distress) stated that they would like to have more vendors. Some markets have begun training vendors who may be currently involved in a different type of farming, or who have the potential of producing food through a market garden. One issue that is exacerbating the shortage, in the Willamette Valley in particular, is the draw of vendors to markets in the Portland area. Apparently attracted there by higher potential profits, vendors are bypassing more local markets to make the trek to Portland each weekend. The drain of vendors may be decreasing the size of other Willamette Valley markets and making things more difficult for markets in small communities.
- However, even in the Portland area, neighborhood markets suffer at times from a decrease or stagnation in their number of vendors because of what they refer to as the appearance of "destination markets." According to these market managers, consumers seem willing to travel a greater distance (sometimes away from their neighborhood market) to shop at the several very large markets that offer the widest product mix, most entertainment, and best atmosphere. The intertwined effects of the shortage of farmer vendors; the potentially evolving tastes of urban farmers' market

customers; and the changing fortunes of individual markets, either suffering a shortage of vendors or elevating to new levels of breadth of products offered, are a fascinating area, and one to watch in the future.

Farmers' Markets and "Community"

A theme that repeatedly emerged during this research is the role of "community" in farmers' markets. It appeared at the beginning of the project with observations by the Project Advisory Committee. They perceived a shift in the purpose for forming farmers' markets over time from being primarily a market channel for farmers to being a tool to build community. These observations were operationalized in the survey of farmers' market managers. The data from the survey revealed the important role community building had as a catalyst for the growth of new markets.

"Community" emerged again in the three manager-determined elements of good farmers' markets: atmosphere, product, and community. In this instance, market managers indicated a web of connections between communities and farmers' markets. Recruiting and maintaining a supportive customer base is one of the main tasks of market organizers. Customers are enticed to markets by food and socializing, and indeed part of the atmosphere of farmers' markets is the ability for customers to linger and socialize. Customers are subtly educated about the value of the market to the community through their interactions with the farmers who provide the food. However, managers also make concerted efforts to weave their markets into the fabric of their communities in other ways, by fostering relationships with community organizations, local decision makers, and associated businesses for example. Andreatta (2005) described farmers' markets as a location that creates community through its use by farmers *and* customers. To further her point, she used the nonposessive form *farmers market* to indicate the site is not owned by farmers but is a shared place.

Gender may be a parallel theme to the prominence of community in this research. The overwhelming majority of farmers' market managers are women. In addition, the majority of volunteer market managers are women. These women, more often than men, are drawn to volunteering their time or working for a low salary, as is often the case, to start and operate farmers' markets. Researchers have reported a prominence of women associated with community-supported agriculture, and that women had a greater interest in community-building aspects of CSA (DeLind & Ferguson, 1999). Perhaps in a similar manner, farmers' markets are "a space more conducive to women's self-identification and reproductive roles, including community building" (DeLind & Ferguson, p. 197).

The View in a Political Ecology Framework

Political ecology provided a basic framework for inquiry and analysis of the data for this research, as well as the visual model of market management ecology. It is used here to provide a framework for a summary of the findings. Recall for a moment my broadly defined and pragmatic approach to political ecology, consisting of the following fundamental components:

1. A multilevel analysis of influences (local to global) from the natural environment (biophysical) and the political environment (social relations, economics, and policy, which are mediated by power hierarchies), incorporating both contemporary and historic perspectives, was used.
2. Actions of participants result in adaptations or adjustments to influences from the natural and political environments. Decision making plays a crucial role in action.
3. With these two points in mind, this investigation focused on the impacts and adaptations at the local level, in order to aid learning (i.e., curriculum and outreach) or potential replication by others facing similar circumstances.

These pragmatic fundamentals worked well with the goals of this study. The approach assisted in the examination of farmers' markets in the context of both their natural and political environments, and it revealed adaptations to influences from these environments. The differences in how individual farmers' markets are managed, and even their differences in appearance from one another, reflect the socioeconomic situation in a community, its local subculture, local government regulations, and local agro-ecozone influences on season and production of food. National and global influences from regulations and trade agreements, as well as positive public policy such as the Senior Farmers' Market Nutrition Program, all demand a response from market managers.

Highlighting the interplay between theory and application, I used the fundamentals of political ecology as the basis for the visual model of market management ecology. It becomes a means for organizations to see the big picture and to potentially make decisions that maintain their long-term viability. As noted previously, the initial response of market managers to the visual model as a teaching tool was positive. Recall one manager's comment: "I feel encouraged by seeing what I do have an influence over. And, I want it produced as a poster for my board [of directors]! This information could have a lot of meaning to farmers' market boards and opportunities for educating boards and guiding market development or market resources."

Farmers' Market Interactions With the Local Physical Environment

The analysis of Oregon's farmers' markets indicates both a diversity of adaptation and some common elements. Farmers' market vendors are most commonly farmers (producers of food or agricultural products), but some markets allow, and even encourage, craft vendors. The number of vendors varies greatly between markets, with some markets having very small numbers and others rather large numbers. The markets reflect their communities, and often markets that are only a few miles apart will appear and "feel" different, and be managed in different ways. This reflects local subculture, socioeconomics, the characteristics of their organizers, and other local conditions.

The markets are transient in character. They are able to set up quickly and transform a stark street or parking lot into an active marketplace, then dismantle just as quickly. In fact, the physical environment is intrinsic to the nature of farmers' markets. Farmers' markets interact directly with the physical environment in two ways: through the market site and through the products for sale at the market. Unlike retail food venues that operate daily and have a wide selection of food regardless of the season, farmers' markets operate in a localized physical environment exposed to a variety of climatic whims.

The open-air form of farmers' markets, along with requirements that their products be grown locally, anchor farmers' markets to their natural environment. Although these features have established markets as a popular niche in the greater consumer marketplace, they also mean markets require continuous adaptation due to their exposure to the elements and dependence on local farming conditions to provide product (food). The length of the market operating season and times of operation are influenced by the physical comfort for vendors and customers and the seasonal quantity of daylight. Moreover, the influence of the local agro-ecozone conditions impacts on market operations in terms of crop availability. Driven to some extent by the profitability of this market channel, farmers are extending their growing season through the use of low-tech methods. This technological adaptation allows them to farm certain crops nearly all year in some regions, while allowing farming in other regions at times of the year when it would otherwise be too risky. The resulting increase in product availability is lengthening the operating season of some farmers' markets.

Rules established by farmers' markets that require food to be produced within a delineated area tie participating community members to their natural environment by focusing product choices on foods that can be grown in their region. The products sold at farmers' markets change with the season, progressing from hardy and tender greens in the late spring to berries, melons, and tomatoes in midsummer, and finally to apples, nuts, and winter squash in the fall. Food sold in this manner embeds community members in the cycle of seasons through the procession of foods available as the growing season advances. Through participation in the seasonal cycle of food production, community members experience, along with farmers, individual years of low and high productivity that result in scarcity and abundance of specific products at the market.

Historic Influences on Expansion and Decline

Because we are currently in an era of rapid growth of farmers' markets, it is useful to acknowledge that there have been other periods of expansion and decline in numbers of farmers' markets. A historical perspective shows that periods of farmers' market expansion and decline have been associated with major external influences at the greater community, national, and international levels. These major external influences are often part of high-level political economic struggles, with their effects cascading down to local levels. The effects are often beyond the scope of the management solutions (adaptations) of the market organizers and their customers. National economic downturns, wars, demographic shifts, local power hierarchies, lifestyle trends, political activism, and public policy have all had negative and positive influences on farmers' markets. This historical perspective, showing that farmers' market growth and decline have occurred in relation to broader historic processes, warns that processes that enhance or weaken markets will likely occur in the future.

Four surges in the numbers of farmers' markets have occurred during the 20th century (Brown, 2002). These included periods following World War I, during the Great Depression, the 1970s (when the surge was related to political activism and a growing resistance to the dominant food system), and the current period of expansion beginning in the 1990s. Periods of decline include the decades of 1950s and 1960s, and the late 1980s. Mimicking biological systems, even after more than 20 years of decline, a remnant population of markets was extant in the United States in the late 1960s. This population provided a platform from which the markets of the 1970s revival grew. The most recent expansion of farmers' markets began during the 1990s and continues today. The suggestions regarding influences stimulating the current growth in numbers of markets, although plausible, are for the most part speculation. At least one reason for the expansion of the 1990s was the use of farmers' markets to build community—a positive external trend influencing both small and large cities—and farmers' markets continue to adapt, blending community building and farmer income as goals.

Adaptation Through Management Organization

The analysis of market size and management tools revealed that as markets increase in size there is an increase in the number of management tools employed and in management complexity. Each market creates internal structures within its available resources so it can face external challenges and barriers. Each market size category, along a continuum from Micro to Large, has unique characteristics.

Micro markets are very informal and are likely to have a volunteer manager. Because of the low number of vendors, many markets of this size do not use the site-management tools used by larger markets. Only several hours per week are devoted to managing these markets during the market season. Virtually none use any of the organizational structures (such as boards of directors, bylaws, etc.) which are associated with larger markets,

Small markets are slightly more complex than Micro markets and show many signs of more formal organization. They are equally split between having paid managers and having volunteer managers. They are large enough to use rules or guidelines, and they arrange vendors to meet specific goals. They have substantially more vendors than Micro markets, but they use only a few additional hours each week for management, both during the market season and the off-season.

The research identified the transition from being a Small market to being a Medium market as a critical point in a market's growth. There are significant differences in organizational structure between the two size categories that must be taken into account in order to successfully accomplish this transition. Medium markets use a paid manager in all instances; some use additional part-time employees and the more formalized structures of a board of directors, bylaws, and more detailed market rules. A Medium market requires a significantly higher number of hours to manage than a Small market.

Although there are many similarities between Medium and Large markets, there are major differences between these categories with regard to the complexity of labor and resources. The number of hours worked per week by the market managers, both during the market season and the off-season, is higher for Large markets than for Medium markets. Market managers also described qualitative differences between these market size categories.

Variation in Access to Resources

Although farmers' markets are currently growing rapidly, the failure rate is high. The life span of failed markets is brief with nearly half closing after only one season. Older markets fail too. The turnover rate for market managers is also high. These research findings reveal the struggle many markets experience as they attempt to provide a sustainable management organization.

Adequate financial resources are crucial in order for markets to obtain other important resources, including labor hired to perform functions that help grow and sustain markets. There are significant differences in the ability of farmers' markets to obtain financial resources. Market administrative revenue is impressive on an aggregate basis but is unevenly distributed among markets. A small number of markets take in nearly half the total administrative revenue for all markets. Labor resource type (paid or volunteer) is connected to the ability of the market to generate administrative revenue.

Larger markets not only take in more administrative revenue based on their size (number of vendors), but they also charge higher stall fees than smaller markets, adding further to their resources. Smaller markets often experience a circular condition in which they cannot attract sufficient customers because they do not have sufficient vendors, but cannot attract sufficient vendors because they do not have sufficient customers to do so. These markets find it difficult or even impossible to garner the resources necessary to support a market manager whose duty it is to recruit vendors and customers and to handle other important operational aspects of the market. From the data available, five interconnected factors were identified in markets that fail: small size, a high need for products, low administrative revenue, a volunteer manager or manager paid a low salary, and high manager turnover. Additional concerns related to market failure include a correlation between new markets and inexperienced managers, effort thresholds for volunteer managers, and the erratic management structures of Small markets, and Small markets' structural readiness for growth.

Continuous Adaptation and "Atmosphere," "Products," and "Community"

A synthesis of data supported the creation of a visual model of farmers' market management ecology. The ecology of farmers' markets includes extensive external influences and identifies common adaptations through management to attain market goals. A simple conceptualization, the visual model acknowledges external conditions that influence farmers' markets from global to local levels. Market managers identified *atmosphere*, *product*, and *community* as key elements of good farmers' markets. These elements were used as targets or goals for market organizers. The model represents farmers' markets in an environment made up of influences from natural and political conditions at the international, national, and local levels. Markets adapt to these conditions through management, and their adaptations are visible; they create an atmosphere conducive to socializing and sales, they procure a variety of high-quality products, and they build community support through the creation of a loyal customer base and integration into local social and economic spheres. Much of the ability to excel in the key traits of successful farmers' markets is based on the use of management tools. Individual farmers' markets have access to varying numbers of people and quantities of time and revenue, and this situation impacts the ability to manage, and therefore, the level of management markets receive.

Recommendations

Discussed here are four broad, summarizing recommendations, as well as one set of very specific recommendations. The findings from this research support recommendations that can be implemented by communities that are planning on starting a market or individual markets that are already operating. However, it is important to keep in mind that equally important are permanent, sustainable solutions based on monitoring and changing state and federal policy to ensure continued enhancement of farmers' markets. Recommendations for attaining this type of change are more abstract.

Recommendation 1:
Plan New Markets Carefully to Ensure Success

Market organizers should spend a considerable length of time deciding whether, and how, to open a new market. Improved planning and promotion prior to opening a market may help with some of the issues confronting it in its first year of operation. An important part of the planning process is setting a goal for the size of the market in general, or by year, so that resources can match the scale of the market. Planning for size is the first step in creating a viable management organization that will endure growth in size, as well as challenges and conflicts—more detail on this are in recommendation 2. Market size will be influenced by community population density, population subculture (interest in purchasing local food and in the experience of open-air markets), and other factors influencing the scale of a market from the demand side. Local farm settlement patterns (number and type of small farms), agro-ecozone (soil types, climate, etc.), and other factors influence the scale of the market from the supply side. Organizers should carefully assess whether there is sufficient customers to support a market and whether there are sufficient farmers and/or market gardeners to support a market.

Training programs for new managers and boards of directors may reduce the number of markets that close after a brief life and the level of manager turnover in operating markets. Recommendation 5 addresses applied research and outreach.

Markets in some communities will always have difficulty generating enough administrative revenue to support a paid manager. Some potential solutions for this issue are outlined in recommendation 3.

These recommendations apply to newer markets. Remember, two of the nine markets that closed following the 2002 manager survey were older markets. The conditions associated with these markets are far more complex and require further investigation.

Recommendation 2: Management Organization Should Be Appropriate for the Size of the Market

As indicated in recommendation 1, planning for the size of a market is an important step in creating an organization that will have the skill resources and financial resources for long-term operation. The research findings on

TABLE 7.1. Summary of management recommendations for market size categories.

Management Tool	Micro (5–8)	Small (9–30)	Medium (31–55)	Large (56–90)
Volunteer manager	😐	☺	💣	💣
Paid manager	😐	☺	☺	☺
Design market layout	O	☺	☺	☺
Market guidelines	☺	☺	☺	☺
Site map	O	☺	☺	☺
Arrange vendors to meet goals	😐	☺	☺	☺
Board of directors	😐	☺	☺	☺
Bylaws	O	☺	☺	☺
Board officers	O	O	☺	☺
Board committees	O	O	O	☺
Additional employees	O	😐	☺	☺
Plan for minimum hours for a manager per week per market in-season	7	10+	20+	30+
Plan for minimum hours for a manager per week per market off-season	2+	3+	10+	15+
Plan for minimum additional employee hours per week per market	0	4+	20+	32+

O = Not necessary; ☺ = Necessary; 😐 = Useful; 💣 = Dangerous.

market structure identified an association between markets of specific sizes and the management tools they used. Table 7.1 summarizes recommendations for farmers' markets based on the four size categories used in this study. The icons represent whether a management tool is "not necessary," "useful," "necessary," or "dangerous." Navigating changes between size categories is not easy. For instance, a market's financial situation may challenge its growth—there may be a need for a paid manager, but the market may not be able to afford one. Funding limitations that constrain additional hours being worked during the market season and off-season may make a market less efficient or threaten its stability.

Recommendation 3: Focus Market Resources on the Local Market and Focus Collective Resources to Address State and Federal Policy

Farmers' market management (managers, boards of directors, organizers, and vendors) have limited resources in terms of funds and staff. These resources must be directed to address issues they can affect. Based on the findings of this research, individual managers should focus on the three core traits of successful markets: atmosphere, products, and community. Individual market organizations may not have any impact on higher-level (state, national, and international) issues and would therefore waste precious resources trying to. This recommendation to devote resources to internal issues is not a recommendation to become insular. The community of farmers' market managers is also part of the ecology of farmers' markets. Regulatory barriers are an issue for farmers' markets, but are perhaps not the best use of time for an individual manager. As a group, however, these issues can be engaged and change can be realized. This is one of the reasons for forming statewide farmers' market associations.

Collective action of farmers' markets can impact policy beyond the local level. With a statewide organization, sentinel managers can inform other managers of state and national issues and apply group pressure on state agencies. Even the appearance of a statewide group garners respect by state agencies. However, statewide organizations are difficult to sustain. Because they are often overseen by a small, dedicated group of volunteers made up of market managers, burnout is a real possibility. A salaried "executive director" is needed to handle day-to-day operations of any such organization and to act as its representative to statewide government agencies and individual communities. The success of instituting membership dues for individual farmers' market organizations may provide a small operating budget. However, because most farmers' markets struggle to pay their own managers, it is unlikely that dues will ever support an operating budget sufficient to provide salaried staff for statewide organizations.

National policy needs to be monitored and influenced by advocates of farmers' markets as well. There is a nascent national organization of farmers' markets called the Farmers' Market Coalition that plans to fill this role. However, as with state-level organizations, financing the support of national-level advocacy organizations is a problem. Many markets are challenged to meet their own operating costs and are not in a position to send sufficient dues upstream to maintain state- and national-level organizations. The eventual resolution may have to come from schemes such as a tax on an item for which farmers' markets represent an positive alternative, in the same way that cigarette taxes are often proposed to fund some healthcare programs.

Recommendation 4:
Some Markets Should Pursue Community Financial Support

Markets in some communities will always have difficulty generating enough administrative revenue to support a paid manager and other important market operations. Because of what I described as the circular condition of smaller farmers' markets, some markets may be viable only through financial or labor resources provided through civic or government entities. There are precedents for this in Oregon. The market manager survey indicated that seven markets were connected with downtown associations or city governments. These organizations support their farmers' markets by providing a salaried staff member for management and other amenities. If a farmers' market is seen as convening an important part of a local economy and enhancing the community's quality of life, there might be more government and economic-development sector interest in supporting it.

Faith, environmental, and health organizations are other potential community sponsors. For instance, the Ecumenical Ministries of Oregon have demonstrated their interest in community food security through production of local food guides and a recent grant-funded effort to link church members with local food production through community-supported agriculture and farmers' market coupons. Many environmental organizations identify locally produced food as good for the environment but are not likely directly involved in the support of individual farmers' markets. The current support of farmers' markets in Oregon and other states by health-care provider Kaiser Permanente offers a model for private-sector support that holds some potential for providing start-up resources for markets in a number of communities. Seeking this type of support from a variety of organizations may be the only option for a stable market in communities that do not have the population base to generate sufficient administrative revenue to support a manager and other basic organizational amenities.

Access to financial resources is an issue related to national policy with significant impacts on farmers' markets, particularly small ones. Small markets are expected to be self-sustaining, while other services provided by the public and the community do not have similar expectations. Public funds support services that enhance the global trade of food products, but a similar level of support is not available to support local agricultural markets. That is a political decision and as Lang (1999) pointed out, food systems are the outcome of policy and political choices.

Recommendation 5: Applied Research and Outreach Are Necessary for the Success of Farmers' Markets

There is a need for continued applied research supporting farmers' markets and educational outreach to managers, boards of directors, and vendors. Applied research has provided and will continue to provide markets, individually and collectively, with the information they need to impact public policy. Applied research can also assist markets in making strategic management decisions. The research findings reveal potentially key components of an educational outreach curriculum. For instance, two research findings offer an opportunity for educational enhancement. The first is regarding directing resources to the key traits of good farmers' markets. The second is the notable underestimation by managers of the external influences on their markets. Together these can be key components of an educational outreach curriculum using the market visual model to assist market organizers in addressing strategic planning and resource allocation. Furthermore, training in farmers' market management, farmers' market-based business management, concepts of marketing, innovative crop production techniques, and other subject areas may assist managers and vendors in maintaining their edge on more commercialized businesses that may appropriate products, appearances, and techniques.

The analysis and recommendations presented here show the effectiveness of using a pragmatic political ecology framework. This is one step in moving concepts of a complex theory (political ecology) to application by non-academics in real-world situations. This type of outreach is long overdue from the social sciences. It packages a way of analyzing complex systems for use by real people. Recalling the utility of political ecology in applied research discussed previously, Robbins' (2004) concept of political ecology as "hatchet" or "seed" is an important theme. Embracing the "seed" category, this investigation uses political ecology to provide farmers' market organizers a view of their goals, the challenges and barriers to reaching their goals from multiple external levels, and a look at how market organizers and practitioners who work with them successfully adapt to the challenges and barriers they encounter. This develops political ecology beyond critique into being a problem-solving tool for use within localized organizations.

Appendix

Research Methods

The research methods used and the approach to the analysis of the data reflect the pragmatism of applied research. The data are primarily from farmers' market managers, with additional data from members of farmers' market boards of directors (farmer vendors and community members). Consequently, the work represents a market manager/management perspective.

Research Design

The data were collected using both qualitative and quantitative methods. These methods have often been intertwined in anthropological research. For instance, H. Russell Bernard's volumes on methods in anthropology and social research (1995, 1998, 2000) include guidance on both approaches. This approach is also referred to as a *mixed methods approach* (Cresswell, 2003) and is routinely utilized by other social science disciplines.

The qualitative and quantitative methods were used in an iterative manner, with each approach guiding and complementing the other. Using mixed methods triangulates data, "seeking convergence across qualitative and quantitative methods" (Trotter & Schensul, 1998, p. 719), and enhances the reliability and validity of the data (Cresswell, 2003). Qualitative methods utilized in this study included participant observation, interviews, and focus groups. Quantitative methods included a survey questionnaire administered by telephone.

A goal of this research is to produce information that is both useful to and used by farmers' market managers and boards of directors, as well as academics and food advocates. By "useful" I mean the issue being addressed by the research is a priority for the participants. By "used" I mean it is applied to individual circumstances—ideally, with positive results.

As a facet of this applied approach, this study also incorporated a participatory orientation. Participatory approaches have deep roots in several traditions. According to Chambers (1994), participatory methods in agricultural development grew from activist participatory research associated with adult education and the work of Paulo Freire (1970); applied anthropology and its focus on the emic perspective, indigenous knowledge, importance of rapport, and other anthropological concepts; as well as farming systems research; and early rapid appraisal techniques such as Rapid Rural Appraisal. According to Sims and Bentley (2002), "Although there are variations on participatory methods, in general, participatory research starts with the concept of collaboration between scientists and farmers as colleagues.... The collaboration resides in the application of different perspectives, knowledge and styles of both parties" (p. 35). In contrasting conventional and participatory approaches to research, "the most important distinctions centre on how and by whom is the research question formulated and by and for whom are research findings used" (Cornwall & Jewkes, 1995, p. 1688).

The participatory approaches informing this study include examples such as Rhoades' Latin American work (2001), as well as domestic research in the United States related to sustainable agriculture (Wuest, McCool, Miller, & Veseth, 1999). Rhoades' work at the International Potato Center in Peru led to the development of the "farmer-back-to-farmer" model emphasizing the emic view of farmers and their participation on interdisciplinary teams. As Wuest, McCool, Miller, and Veseth (1999) noted, participatory research includes the participant as an intellectual partner in the research. In such instances in agricultural development, farmers set the research agenda, contribute indigenous/local knowledge, and often, carry out research on their farms. Participatory approaches of this kind are not a reflexive methodology (Richie, 2002).

Research referred to as participatory may be only slightly participatory, or it may be fully so. There are contextual reasons for selecting approaches that are more or less participatory, in order to address specific types of research and to play to the strengths of both the participants and the researcher. For this project, the participants had an intense interest in the research topic but did not have the time to carry out the research in a fully participatory manner—they did, after all, have part- or full-time jobs managing farmers' markets. Therefore, a collaborative relationship with me as researcher was an efficient match. I used my expertise to design the research protocol, and contributed my time to conduct the survey research and organize the analysis of the quantitative and qualitative data. The participants contributed their knowledge as experts in farmers' market management to design research questions and provide insights on data interpretation and synthesis. Consequently, my roles were researcher, analyst, and compiler; and the roles of the participants were advisor, consultant, analyst, and reviewer. This type of partnership between participants and researchers produces tangible, useful results.

A participatory orientation generally used with farmers was built into this study concentrating on farmers' markets; specifically, a committee of three participants (two farmers' market managers and the president of the Oregon Farmers' Market Association) comprised the Project Advisory Committee for this study. The use of farmer committees to guide research projects is noted by Van de Fliert and Braun (2002) as being among many sound approaches. The Project Advisory Committee assisted in setting a research agenda, contributed to the design of the survey questionnaire, provided consultation and guidance for interpretation and synthesis of the data, and reviewed and critiqued the findings. An additional participatory feature was the access to the additional knowledge and guidance of the market managers who participated in the survey as "subjects," through their additional participation as "researchers." Through a series of focus groups (discussed in more detail later), their input was gathered regarding how I interpreted the data. The Project Advisory Committee was then used to review and critique a draft of the final study findings. This research project changed both me, as a researcher, and the participants, via a continuous process of communication and interaction while we were undertaking planning, implementation, data analysis, interpretation, and synthesis.

The research was conducted under a protocol approved by the institutional review board of Oregon State University for the protection of human subjects. Pseudonyms were used where confidentiality is required within the research protocol. Real names of people, places, and markets are used where the information is a matter of public record.

Study Context

This research focuses on the 53 farmers' markets operating in Oregon during 2002 that had operated during the 2001 season,[1] and the 43 individuals who managed them. The participants in this study were dispersed geographically over 50 sites around the state of Oregon. Furthermore, the markets were dispersed chronologically, with most operating only one day per week for perhaps 30 weeks per year or fewer. Most of the effort of managing a farmers' market takes place after-hours and off-season, via telephone and e-mail. The "community" of farmers' market managers exists via informal telephone networks, an e-mail discussion list (Oregon Farmers' Markets-List or ORFM-L), and the activities of the Oregon Farmers' Market Association (OFMA).

Because of the conditions described previously, "fieldwork" for this study occurred at specific farmers' markets on specific days, or through telephone conversations, e-mail exchanges, and face-to-face contact by formal arrangement or casual contact. The research methods were selected and adapted to meet these conditions.

Research Protocol

Research activities were conducted beginning in January 2002 and continuing until December 2005. A number of methods were used to conduct this research, including: participant observation, a survey administered by telephone, interviews, focus groups, and analysis of public documents.

Participation and Observation

Observations of markets were conducted using in-depth approaches involving the intensive collection of a wide range of data at particular markets. These sessions were prearranged for a specific day with the market managers of the individual markets. Methods used included interviewing and shadowing the market manager, interviewing board members and vendors, mapping and assessing the layout of the market, visiting associated neighborhoods and business areas, and observing market customer and vendor behavior (interactions, conversations, and shopping behavior). This latter method emphasizes what Bernard (2000) referred to as "unobtrusive continuous monitoring" (p. 376), wherein market activities and interactions between participants were observed and recorded without the participants' knowledge. In this way, I was able to observe the market in its "natural" state.

Monitoring the Oregon Farmers' Market e-mail discussion list provided an additional view of farmers' market managers. Interactions on the discussion list include queries from new and established managers regarding how specific situations are handled at other markets; searches for products; news related to legislation, federal policy, or regulatory issues; and debates related to market site management, including whether dogs should be allowed in markets (a controversial issue). The discussion list, for which I act as moderator, is hosted by Oregon State University.

Survey of Participants by Telephone

Because the project called for detailed information for *each* market in Oregon, the best method for obtaining this information was a survey of each farmers' market manager administered by telephone. Like a questionnaire administered by mail, this method of data collection offers the strength of obtaining specific quantitative data. A telephone survey additionally offers the opportunity for extended answers to open-ended questions and clarification of information for both the interviewer and the participant. A telephone survey of this type is very similar to a structured interview.

Creation of the survey questionnaire was carried out during two meetings with the Project Advisory Committee in 2002. The committee formulated many of the research questions during this process including questions about the community-building aspects of farmers' markets, their fee structures, and their sources of revenue. I relied on this process to identify market-site amenities, typical products, and common management structures. Salant and Dillman's work (1994) was consulted for questionnaire design, guidance on formulating questions, and managing the implementation of the survey. The questionnaire was pretested with market managers by telephone. The survey of market managers was conducted between April and June 2002. The final version of the questionnaire was composed of 68 open-ended and closed questions. The survey collected a variety of information on two distinct elements: the market and the market manager. Specific information related to the market (e.g., market revenue and number of vendors) was based on the 2001 season. Answers to other questions reflected the attitudes and opinions of the managers during the spring of 2002. The surveys were administered by myself and a faculty research assistant employed under the project grant.

Of the 53 farmers' markets operating during 2002 that had operated during 2001, 50 participated in the survey. This response rate of slightly more than 94% is gratifying but also crucial for the validity of a research study focusing on a numerically small population.

The three markets that did not participate in the survey were located in the Willamette Valley and comprised Small markets, a category well represented in the rest of the data set, both in terms of geography and scale. These markets were contacted several times.

Interviews of Participants

Bernard (2000, pp. 190–194) outlined a range of interview styles based on a continuum of degrees of control the researcher exercises over participants' responses. The interview styles range from informal to structured. Since the telephone survey functioned as a highly structured interview, I relied principally on informal and unstructured interviews to enhance and augment data. Interviews were a fundamental part of the research and were conducted throughout the project. Interviews during the early phase of the research were aimed at moving my knowledge up the learning curve: I asked questions so I would know what questions to ask. Following the survey of managers and the early analysis of the data, interviews were aimed at interpretation of the data from emic views and were intended to obtain useful examples of management behavior, relevant stories, and so forth from the individual market perspective. These interviews were used strategically to address specific issues I was encountering in the data analysis. Approximately 50 interviews with 30 individuals are recorded in my field notes.

Many of the interviews were conducted with a specific market's manager and members of its board of directors while it market was operating. Other interviews were conducted, by telephone or in person, in a variety of settings. Most interviews were with key participants and could last for hours at a time. Because of the nature of my relationship with the participants, innumerable additional small conversations yielded important bits of information.

Focus Groups

Following a preliminary analysis of the quantitative and qualitative data, six focus groups of market managers and board members were organized and conducted. A total of 29 individuals participated, representing 33 farmers' markets. The groups were organized according to the same regions utilized in the data analysis.

The focus groups were designed to act as a two-way information channel, first by presenting the information from the research back to the participants, then by probing the groups for deeper interpretation of the data to get at the *how* and *why* (Bernard, 2000). Each group was asked a basic set of questions, but discussions were encouraged to follow the interests of the participants. Since the groups were organized according to region, each group interpreted the data from its regional perspectives, a result that would have been difficult to achieve with focus groups of mixed regions. A research assistant recorded detailed notes on a computer for each session. Focus-group input assisted in modifying some segments of the research, and the elaborations of issues offered by participants enrich the data presented in this document.

A final participatory step involved utilizing a small committee of participants to review and critique the research findings. This group was composed of four individuals, including the three Project Advisory Committee members, as well as an additional market manager. The additional participant was added so the group would include individuals who had direct experience with each of the four market sizes utilized in the data analysis.

The review and critique process involved two steps: an initial review of a draft manuscript by individual group members, followed by a meeting of the four reviewers to discuss the findings as a group. The reviewers were asked to perform at least three tasks (as well as providing any additional input they wished to include):

- To correct inaccuracies
- To discuss usefulness and potential applications of the information
- To honestly point out instances where assumptions or interpretations were erroneous

The reviewers returned the materials, and I reviewed and summarized their written comments. For the second step, a meeting of the reviewers was conducted to discuss the material and take advantage of the creative and insightful synergy that takes place in a group setting. In addition to the four reviewers and myself, a research assistant recorded detailed notes on a computer during the meeting. Information from the review process was utilized to improve the quality of information included in this document and to provide a "forum" for points of view on the value of the research, its usefulness, and its potential applications.

Other Important Data Resources

Directories of Oregon farmers' markets for 1998 to 2005 were used as a crucial secondary data source. The directories are published by the Oregon Farmers' Market Association and the Oregon Department of Agriculture. The directories list the markets operating for each year of the publication. These directories were used to identify the year markets began operating and ceased operating. This was accomplished by recording their the first appearance of a market in the directories, which was taken to be the year it opened, and recording the year when it no longer appeared, which was taken to indicate that closure had occurred sometime since the previous season. Since the information in the directories is not perfect, dates were occasionally adjusted when data from direct experience with a market or the survey of market managers indicated a market had opened or closed during a different year.

Data Management and Analysis

Quantitative data from the survey questionnaire were organized and analyzed using Statistical Package for the Social Sciences (SPSS) version 11.5 for Microsoft Windows. Data analysis consisted of descriptive statistics, analysis of variance, and correlation analysis (ANOVA, *t* test, correlation coefficient). Statistical analysis was conducted with consultative support from the Survey Research Center at Oregon State University. Interview data were reviewed and segments were grouped into categories in order to link them with quantitative data from the survey.

One open-ended question from the survey of market managers was particularly important to this research project. The question concerned how market managers defined a "good" farmers' market. Since each manager expressed the answer in many different ways, a *grounded theory* approach was used to analyze the content of answers to this question.

"Grounded theory is an iterative process by which the analyst becomes more and more 'grounded' in the data and develops increasingly richer concepts and models of how the phenomenon being studied really works" (Bernard & Ryan, 1998, p. 608). Grounded theory is a useful technique for identifying categories and concepts that emerge from text, and it helps the researcher link the concepts to theory (Bernard & Ryan). This information was used to create the visual model of market ecology. Influences for this model include "Julian Steward's cultural ecology" (Robbins, 2004, p. 31), placemaking tools by Project for Public Spaces (http:// www.pps.org/info/placemakingtools/), and other ecosystem models.

Creating and Validating Farmers' Market Size Categories

The process of creating farmers' market categories involved examining and integrating both quantitative and qualitative data, and interacting with farmers' market managers for guidance and validation. Beginning with my own sense of market scale and the nature of markets in a broad sense, I examined the frequency distribution for number of vendors, and identified breaks in the distribution that could potentially form size-category boundaries. I then developed a preliminary set of categories and discussed them, on an individual basis, with members of the Project Advisory Committee. The committee offered adjustments based on their perceptions of market size categories. Most of these were integrated into a working set of categories, although there was not exact agreement among the committee members.

The working set of categories was then examined statistically. Because market revenue is highly correlated with market size ($r(50) = 0.835$, $p < 0.0001$), it was used to validate the categories. A one-way analysis of variance (ANOVA) showed that the relationship between the market size categories and market revenue is highly significant ($F(3, 46) = 41.30$, $p < 0.0001$). Tukey's HSD test was used to test for significant differences between categories within the one-way ANOVA. The test revealed that the categories were significantly different for Small, Medium, and Large markets ($p < 0.05$). It was not, however, significantly different between

Micro and Small markets ($p > 0.05$). This notwithstanding, there are sufficient qualitative differences between Micro and Small markets in terms of type of site, management structures, and other factors that the distinction between the two categories is useful. The market size categories were reviewed and validated by 29 market managers and board members during the six focus-group sessions. One advisory committee member felt that the upper end of the small category was too large, but all other focus-group participants endorsed the categories. This extensive process has rendered a reliable set of categories.

ENDNOTE

1. Collecting data from markets that had operated for a full market season was a key component of the research. Although there were 61 farmers' markets operating in 2002, only 53 had operated during 2001.

References

Adrian, J. (1982). Farmers' markets in Alabama. *Highlights of Agricultural Research, 29*(2), 16.

American Farmland Trust. (2001). *Protecting our most valuable resources: Results of a national public opinion poll* [Online report]. Retrieved July 17, 2005, from http://www.farmland.org/news_2001/survey_summary_statistics.pdf

Anderson, J. (1995). *Regional farmers' market: A marketing and design study conducted for Springfield and Columbia, MO.* Washington, DC: U.S. Dept. of Agriculture, Agricultural Marketing Service, Transportation and Marketing Division, District of Columbia.

Andreatta, S. (2000). Marketing strategies and challenges of small-scale organic producers in central North Carolina. *Culture and Agriculture, 22*(3), 40–50.

Andreatta, S. (2005). Urban connections to locally grown produce. In P. Barlett (Ed.), *Urban place: Reconnecting with the natural world* (pp. 117–140). Cambridge, MA: MIT Press.

Andreatta, S., & Wickliffe, W., II. (2002). Managing farmer and consumer expectations: A study of a North Carolina farmers' market. *Human Organization, 61*(2), 167–176.

Archer Daniels Midland. (2004). *What if we looked at the world as one giant farm field?* [Print advertisement]. Retrieved August 2, 2004, from http://www.admworld.com

AVG survey, how big is the roadside market industry? (1977a). *American Vegetable Grower, 25*(2), 30–31, 66–67, 70–71.

Baber, L., & Frongillo, E. (2003). Family and seller interactions in farmers' markets in Upstate New York. *American Journal of Alternative Agriculture, 18*, 87–94.

Beardsworth, A., & Keil, T. (1997). *Sociology on the menu: An invitation to the study of food and society*. London: Routledge.

Beaverton Farmers' Market, Inc. (2005). *2005 vendor handbook.* Retrieved October 27, 2005, from http://www.beavertonfarmers-market.com/bfm6_info.htm/

Bennett, J. (1969). *The northern plainsman: Adaptive strategy and agrarian life.* Chicago: Aldine.

Bennett, J. (1990). Ecosystems, environmentalism, resource conservation and anthropological research. In E. Moran (Ed.), *The ecosystem approach in anthropology, from concept to practice* (pp. 435–458). Ann Arbor, MI: The University of Michigan Press.

Bernard, H. (1995). *Research methods in anthropology, qualitative and quantitative approaches* (2nd ed.). Walnut Creek, CA: Altamira Press.

Bernard, H. (Ed.) (1998). *Handbook of methods in cultural anthropology.* Walnut Creek, CA: Altamira.

Bernard, H. (2000). *Social research methods, qualitative and quantitative approaches.* Thousand Oaks, CA: Sage Publications, Inc.

Bernard, H., & Gery, W. (1998). Text analysis, qualitative and quantitative methods. In H. Bernard (Ed.), *Handbook of methods in cultural anthropology* (pp. 595–646). Walnut Creek, CA: Altamira Press.

Bettis, S. (1969). *Market days, an informal history of the Eugene Producers' Public Market.* Eugene, OR: Lane Pomona Grange Fraternal Society.

Big River. (2006). *Mission statement printed on the menu.* Big River Restaurant, Corvallis, OR.

Blaikie, P., & Brookfield, H. (1987). *Land degradation and society.* London: Methuen.

Blank, S. (1998). *The end of agriculture in the American portfolio.* Westport, CT: Quorum books.

Bonnano, A. (1994). *From Columbus to Conagra: The globalization of agriculture and food.* Lawrence, KS: University Press of Kansas.

Born, B., & Purcell, M. (2006). Avoiding the local trap: Scale and food systems in planning research. *Journal of Planning Education and Research, 26*, 195–207.

Brooker, J., Eastwood, D., & Gray, M. (1993, February). Direct marketing in the 1990s: Tennessee's new farmers' markets. *Journal of Food Distribution Research, 24*(1), 127–138.

Brosius, J. (1999). Green dots, pink hearts: Displacing politics from the Malaysian rain forest. *American Anthropologist, 101*, 36–37.

Brown, A. (2001). Counting farmers markets. *The Geographical Review, 91*(4), 655–674.

Brown, A. (2002). Farmers' market research 1940–2000: An inventory and review. *American Journal of Alternative Agriculture, 17*(4), 167–176.

Burns, A., & Johnson, D. (1996). *Farmers' market survey report.* Washington, DC: U.S. Department of Agriculture, Agriculture Marketing Service, Transportation and Marketing Division, Wholesale and Alternative Markets Program.

Butler, L., & Maronek, D. (2002). *Urban and agricultural communities: Opportunities for common ground* (Task Force Rep. No. 138). Ames, IA: Council for Agricultural Science and Technology (CAST).

Center for Rural Pennsylvania. (2002). *Starting and strengthening farmers' markets in Pennsylvania* (2nd ed.). Harrisburg, PA: The Center for Rural Pennsylvania.

Chambers, R. (1994). Participatory rural appraisal (PRA): Analysis of experience. *World Development, 22*(9), 1253–1268.

Chase, N., & Winn, J. (1981). Farmers' markets: An idea whose time has come…again. *Journal of Extension, 19*(2), 12–17.

Cone, C., & Myhre, A. (2000). Community supported agriculture: A sustainable alternative to industrial agriculture? *Human Organization, 59*, 187–197.

Conner, D. (2004). Expressing values in agricultural markets: An economic policy perspective. *Agriculture and Human Values, 21*, 27–35.

Cornwall, A., & Jewkes, R. (1995). What is participatory research? *Social Science and Medicine, 41*(12), 1667–1676.

Corum, V., Rosenzweig, M., & Gibson, E. (2001). *The new farmers' market: Farm-fresh ideas for producers, managers & consumers.* Auburn, CA: New World Publishing.

Cresswell, J. (2003). *Research design, qualitative, quantitative, and mixed methods approaches* (2nd ed.). Thousand Oaks, CA: Sage Publications.

Davis, V. (1978). *Small farmers and their market: Relic of the past or an option for the future*. Unpublished doctoral dissertation, Syracuse University, Syracuse, NY.

DeLind, L. (1993). Market niches, "Cul de Sacs," and social context: Alternative systems of food production. *Culture and Agriculture, 45/46*, 7–12.

DeLind, L. (2003). Considerably more than vegetables, a lot less than community, the dilemma of community supported agriculture. In J. Adams (Ed.), *Fighting for the farm, rural America transformed* (pp. 192–206). Philadelphia: University of Pennsylvania Press.

DeLind, L., & Ferguson, A. (1999). Is this a women's movement? The relationship of gender to community-supported agriculture in Michigan. *Human Organization, 58*(2), 190–200.

Deweese, P. (1974). *The Eastern farmers' market: An urban ethnography*. Unpublished doctoral dissertation, Purdue University, West Lafayette, IN.

Doorman, F. (1991, fall). A framework for the rapid appraisal of factors that influence the adoption and impact of new agricultural technology. *Human Organization, 50*(3), 235–244.

Eckstein, L., & Plattner, S. (1978). Ethnicity and occupations in Soulard Farmers Market, St. Louis, Missouri. *Urban Anthropology, 7*(4), 361–371.

Ecotrust. (2003). *Section Z: Start a revolution with a tomato*. Portland, OR: Ecostrust, Inc. Retrieved March 5, 2008, from http://www.sectionz.info/ISSUE_3/

Eigo, G., & Engeman, R. (2002). *A market for the city, the history of Portland's public markets.* Portland, OR: Oregon Historical Society.

Estes, E. (1985, June). *Community farmers' markets in North Carolina: A survey of consumers and sellers in 1981* (Bulletin 472, p. 14). Raleigh, NC: North Carolina Agricultural Research Service, North Carolina State University.

Farmers' markets are coming to town! (1977). *American Vegetable Grower, 25*(6), 11–13, 42.

Feenstra, G. (2002). Creating space for sustainable food systems: Lessons from the field. *Agriculture and Human Values, 19*, 99–106.

Feenstra, G., & Lewis, C. (1999). Farmers markets offer new business opportunities for farmers. *California Agriculture, 53*(6), 25–29.

Festing, H. (1998). *Farmers' markets: An American success story*. Bath, U.K.: Eco-logic Books.

Fetters, T., & Meirer, H. (1980). *Direct farm marketing practices and activities in Oregon* (Special Rep. No. 570). Corvallis, OR: Oregon State University Extension Service.

Fisher, A. (1999). *Hot peppers and parking lot peaches: Evaluating farmers' markets in low income communities.* Venice, CA: Community Food Security Coalition.

Friedland, W. (2004). Agrifood globalization and commodity systems. *International Journal of Sociology of Agriculture and Food, 12.* Retrieved March 5, 2008, from http://www.csafe.org.nz/ijsaf/archive/vol12/Friedland.pdf

Friedmann, H. (1994). Distance and durability: Shaky foundations of the world food economy. In P. McMichael (Ed.), *The global restructuring of agro-food systems* (pp. 258–276). New York: Cornell University Press.

Friere, P. (1970). *Pedagogy of the oppressed.* New York: Herder & Herder.

Gillespie, G., Hilchey, D., Hinrichs, C., & Feenstra, G. (2007). Farmers' markets as keystones in rebuilding local and regional food systems. In C. Hinrichs & T. Lyson (Eds.), *Remaking the North American food system, strategies for sustainability* (pp. 65–83). Lincoln, NE: University of Nebraska Press.

Goland, C. (2002). Community supported agriculture, food consumption patterns, and member commitment. *Culture and Agriculture, 24*(1), 14–25.

Goldschmidt, W. (1978). *As you sow: Three studies in the social consequences of agribusiness.* Montclair, NJ: Allanheld, Osmun & Co. Publishers.

Goldschmidt, W. (1979). On the interdependence between utility and theory. In W. Goldschmidt (Ed.), *The uses of anthropology* (pp. 1–13). Washington, DC: American Anthropological Association.

Goldschmidt, W. (1998). The urbanization of rural America. In K. Thu & E. Durrenberger (Eds.), *Pigs, profits and rural communities* (pp. 183–198). New York: State University of New York Press.

Goldschmidt, W. (2001). Notes toward a theory of applied anthropology. Malinowski Award Lecture. *Human Organization, 60*(4), 423–429.

Goodman, A., & Leatherman, T. (1998). Traversing the chasm between biology and culture: An introduction. In A. Goodman & T. Leatherman (Eds.), *Building a new biocultural synthesis* (pp. 3–41). Ann Arbor, MI, University of Michigan Press.

Govindasamy, R., Zurbriggen, M., Italia, J., Adelaja, A., Nitzsche, P., & VanVranken, R. (1998). *Farmers markets: Manager characteristics and factors affecting market organization* (New Jersey Agricultural Experiment Station P-02137-8-98). New Brunswick, NJ: Rutgers, The State University of New Jersey.

Greenberg, J., & Park, T. (1994). Political ecology. *Journal of Political Ecology, 1*, 1–12. Retrieved March 5, 2008, from http://jpe.library.arizona.edu/Volume1/Volume_1_1.html

Grey, M. (2000). The industrial food stream and its alternatives in the United States: An introduction. *Human Organization, 59*(2), 143–150.

Grogan, S., Carruthers, L., Rolle, A., & Hole, M. (1919). *Municipal markets in cities having a population of over 30,000.* Washington, DC: U.S Bureau of the Census.

Guptill, A., & Wilkins, J. (2002). Buying into the food system: Trends in food retailing in the US and implications for local foods. *Agriculture and Human Values, 19*, 39–51.

Hamilton, N. (2002). *Farmers' markets rules, regulations, and opportunities* [An agricultural law research article]. Fayetteville, AR: The National Agricultural Law Center, University of Arkansas.

Headd, B. (2003). Redefining business success: Distinguishing between closure and failure. *Small Business Economics, 21*(1), 51–61.

Heffernan, W., & Constance, D. (1994). Transnational corporations and the globalization of the food system. In A. Bonanno, L. Busch, W. Friedland, L. Gouveia, & E. Mingione (Eds.), *From Columbus to Conagra, the globalization of agriculture and food* (pp. 29–51). Lawrence, KS: University Press of Kansas.

Heisley, D., McGrath, M., & Sherry, J. (1991). To everything there is a season: A photoessay of a farmers' market. *Journal of American Culture, 14*(3), 53–79.

Hendrickson, M. (2005, February). *Direct and niche marketing—Is it here to stay?* Paper presented at the 2005 USDA Agricultural Outlook Forum, Arlington, VA.

Hendrickson, M., & Heffernan, W. (2002). *Concentration of agricultural markets.* Columbia, MO: Department of Rural Sociology, University of Missouri. Retrieved February 28, 2006, from http://www.foodcircles.missouri.edu/CRJanuary02.pdf

Hendrickson, M., & Heffernan, W. (2005). *Concentration of agricultural markets.* Columbia, MO: Department of Rural Sociology, University of Missouri. Retrieved February 27, 2006, from http://www.foodcircles.missouri.edu/CRJanuary05.pdf

Hinrichs, C. (2000). Embeddedness and local food systems: Notes on two types of direct agricultural markets. *Journal of Rural Studies, 16*, 295–303.

Hinrichs, C. (2001). *Observations and concerns of Iowa farmers' market managers* (Summary of research findings). Ames, IA: Department of Sociology, Iowa State University. Retrieved July 2005, from www.soc.iastate.edu/extension/managers_report.pdf

Hinrichs, C., Gillespie, G., & Feenstra, G. (2004). Social learning and innovation at retail farmers' markets. *Rural Sociology, 69*(1), 31–58.

Hodges, L. (1997) *Guides for development of a farmers market.* Lincoln, NE: Cooperative Extension, Institute of Agriculture and Natural Resources, University of Nebraska.

Holt, G. (2005, May). Local foods and local markets: Strategies to grow the local sector in the UK. *Anthropology of Food, 4*, 1–20. Retrieved March 5, 2008, from http://aof.revues.org/document179.html

Hughes, M., & Mattson, R. (1992). *Farmers' markets in Kansas: A profile of vendors and market organization* (Report of Progress No. 658). Manhattan, KS: Agricultural Experiment Station, Kansas State University.

Ikerd, J. (1995, October/November). *The role of marketing in sustainable agriculture.* Paper presented at the 87th Annual Meeting of the American Society of Agronomy, St. Louis, MO. Retrieved March 3, 2006, from http://www.ssu.missouri.edu/faculty/jikerd/papers/Stl-mkt.htm

Ikerd, J. (1998). Sustainable agriculture, rural economic development, and large-scale swine production. In K. Thu & E. Durrenberger (Eds.), *Pigs, profits, and rural communities* (pp. 157–169). Albany, NY: State University of New York Press.

Jansen, K. (1998). *Political ecology, mountain agriculture, and knowledge in Honduras.* Amsterdam: Thela Publishers.

Johnson, D., & Bragg, E. (1994). *1994 National farmers' market directory.* Washington, DC: USDA.

Johnson, D., Lewis, L., & Bragg, E. (1996). *1996 National farmers' market directory.* Washington, DC: USDA.

Kambara, K., & Shelley, C. (2002). *The California agricultural direct marketing study.* Davis, CA: California Institute of Rural Studies.

Kloppenburg, J., Hendrickson, J., & Stevenson, G. (1996). Coming in to the foodshed. *Agriculture and Human Values, 13*, 33–42.

Kottak, C. (1998). Presenting anthropology to diverse audiences. In H. Bernard (Ed.), *Handbook of methods in cultural anthropology* (pp. 737–761). Walnut Creek, CA: Altamira Press.

Kottak, C. (1999). The new ecological anthropology. *American Anthropologist, 100*(1), 23–35.

Kremen, A., Greene, C., & Hanson, J. (2003). *Organic produce, price premiums, and eco-labeling in U.S. farmers' markets* (VGS-301-01). Washington, DC: USDA Economic Research Service.

Lang, T. (1999). The complexities of globalization: The UK as a case study of tensions within the food system and the challenge for food policy. *Agriculture and Human Values, 16*, 169–185.

Lev, L., & Stephenson, G. (1998). *Analyzing three farmers' markets in Albany and Corvallis, Oregon* (Small Farm Tech. Rep. Series No. 2). Corvallis, OR: Oregon State University Extension Service. Retrieved March 5, 2008, from http://smallfarms.oregonstate.edu/sites/default/files/publications/techreports/TechReport2.pdf

Lev, L., & Stephenson, G. (1999). Dot posters: A practical alternative to written questionnaires and oral interviews. *Journal of Extension, 37*(5). Retrieved March 5, 2008, from http://www.joe.org/joe/1999october/tt1.html

Lev, L., & Stephenson, G. (2001). *Rapid assessment of five Oregon farmers' markets: Quantitative results* (Oregon Small Farms Tech. Rep. No. 3). Corvallis, OR: Oregon State University Extension Service. Retrieved March 5, 2008, from http://smallfarms.oregonstate.edu/sites/default/files/publications/techreports/TechReport3.pdf

Lev, L., & Stephenson, G. (2002). A learning approach to strengthening farmers' markets. *Journal of Food Distribution and Research, 33*(1), 112–116.

Lev, L., Stephenson, G., & Brewer, L. (2002). *Hillsdale farmers' market rapid market assessment* (Oregon Small Farms Tech. Rep. No. 13). Corvallis, OR: Oregon State University Extension Service. Retrieved March 10, 2006, from http://smallfarms.oregonstate.edu/publications/techreports/TechReport13.pdf/

Linstrom, H. (1978). *Farmer-to-consumer marketing* (ESCS–01). Washington, DC: U.S. Department of Agriculture, Economics, Statistics, and Cooperatives Service.

Lloyd, R., Nelson, J., & Tilley, D. (1987). *Should I grow fruits and vegetables?* (Oklahoma State University Extension Facts 185). Stillwater, OK: Oklahoma State University.

Lockeretz, W. (1986). Urban consumers' attitudes towards locally grown produce. *American Journal of Alternative Agriculture, 1*(2), 83–88.

Lockwood, W. G. (1972). *Periodic markets: Source materials on markets and fairs in peasant society* (Exchange Bibliography, No. 341). Moticello, IL: Council of Planning Librarians.

Lyson, T. (2004). *Civic agriculture: Reconnecting farm, food, and community*. Medford, MA: Tufts University Press.

Lyson, T. (2007). Civic agriculture and the North American food system. In C. Hinrichs & T. Lyson (Eds.), *Remaking the North American food system, strategies for sustainability* (pp. 19–32). Lincoln, NE: University of Nebraska Press.

Lyson, T., Gillespie, G., Jr., & Hilchey, D. (1995). Farmers' markets and the local community: Bridging the formal and informal economy. *American Journal of Alternative Agriculture, 10*(3), 108–113.

McGrath, M., Sherry, J., & Heisley, D. (1993). An ethnographic study of an urban periodic marketplace: Lessons from the Midville market. *Journal of Retailing, 69*(3), 280–319.

Moran, E. (1990). Ecosystem ecology in biology and anthropology: A critical assessment. In E. Moran (Ed.), *The ecosystem approach in anthropology, from concept to practice* (pp. 3–40). Ann Arbor, MI: The University of Michigan Press.

Moran, E. (2000). *Human adaptability: An introduction to ecological anthropology* (2nd ed.). Boulder, CO: Westview Press.

Murray, M. (2007). *Challenges to re-embedding food systems: Tracing the decline and revival of a farmers' market on the edge*. Unpublished master's thesis, Oregon State University, Corvallis, OR.

Oberholtzer, L., & Grow, S. (2003). *Fresh from the farm: Overview and characteristics of producer-only markets in the mid-Atlantic region.* Arlington, VA: Henry A. Wallace Center for Agricultural and Environmental Policy.

Oregon Department of Agriculture (ODA). (1998). *Oregon's farmers' markets* [Pamphlet]. Salem, OR: Oregon Department of Agriculture.

Oregon Department of Agriculture (ODA). (1999). *Oregon's farmers' markets* [Pamphlet]. Salem, OR: Oregon Department of Agriculture.

Oregon Department of Agriculture (ODA). (2000). *Oregon's farmers' markets* [Pamphlet]. Salem, OR: Oregon Department of Agriculture.

Oregon Department of Agriculture/Oregon Farmers' Market Association (ODA/OFMA). (2001). *Oregon's farmers' markets* [Pamphlet]. Portland, OR: Oregon Farmers' Market Association and Oregon Department of Agriculture.

Oregon Farmers' Market Association/Oregon Department of Agriculture (OFMA/ODA). (2002). *Oregon's farmers' markets* [Pamphlet]. Portland, OR: Oregon Farmers' Market Association and Oregon Department of Agriculture.

Oregon Farmers' Market Association/Oregon Department of Agriculture (OFMA/ODA). (2003). *Oregon's farmers' markets* [Pamphlet]. Portland, OR: Oregon Farmers' Market Association and Oregon Department of Agriculture. Portland, OR.

Oregon Farmers' Market Association/Oregon Department of Agriculture (OFMA/ODA). (2004). *Oregon's farmers' markets* [Pamphlet]. Portland, OR: Oregon Farmers' Market Association and Oregon Department of Agriculture.

Oregon Farmers' Market Association/Oregon Department of Agriculture (OFMA/ODA). (2005). *Oregon's farmers' markets* [Pamphlet]. Portland, OR: Oregon Farmers' Market Association and Oregon Department of Agriculture.

Paulson, S., & Gezon, L. (Eds.). (2005). *Political ecology across spaces, scales, and social groups.* New Brunswick, NJ: Rutgers University Press.

Paulson, S., Gezon, L., & Watts, M. (2003). Locating the political in political ecology: An introduction. *Human Organization, 62*(3), 205–217.

Paulson, S., Gezon, L., & Watts, M. (2005). Politics, ecologies, genealogies. In S. Paulson & L. Gezon (Eds.), *Political ecology across spaces, scales, and social groups* (pp. 17–37). New Brunswick, NJ: Rutgers University Press.

Payne, T. (2002). *U.S. farmers' markets—2000, a study of emerging trends.* Washington, DC: U.S. Department of Agriculture, Agriculture

Marketing Service, Transportation and Marketing Division, Marketing Services Branch.

Pelsue, N., Jr. (1984, October). *Consumers at farmers' markets and roadside stands in Vermont* (Research Rep.: Vermont Agricultural Experiment Station No. 41, p. 19). Burlington, VT: University of Vermont.

Pelto, G., & Vargas, L. (1992). Introduction: Dietary change and nutrition. *Ecology of Food and Nutrition, 27*, 159–161.

Peters, S., Jordan, N., Adamek, M., & Alter, T. (Eds.). (2005). *Engaging campus and community: The practice of public scholarship in the state and land-grant university system.* Dayton, OH: Charles F. Kettering Foundation.

Pirog, R., Van Pelt, T., Enshayan, K., & Cook, E. (2001). *Food, fuel, and freeways: An Iowa perspective on how far food travels, fuel usage, and greenhouse gas emissions.* Ames, IA: Leopold Center for Sustainable Agriculture.

Plattner, S. (1978). Public markets: Functional anachronisms or functional necessities? *Ekistics, 273*, 444–446.

Plattner, S. (1982). Economic decision making in a public marketplace. *American Ethnologist, 9*, 399–420.

Plattner, S. (1984). Economic decision making of marketplace merchants: An ethnographic model. *Human Organization, 43*, 252–264.

Plattner, S. (1989a). Economic behavior in markets. In S. Plattner (Ed.), *Economic anthropology* (pp. 209–221). Stanford, CA: Stanford University Press.

Plattner, S. (1989b). Markets and marketplaces. In S. Plattner (Ed.), *Economic anthropology* (pp. 171–208). Stanford, CA: Stanford University Press.

Portland Farmers' Market. (2004). *Vendor handbook, 2004* (A handbook for vendors). Portland, OR: Portland Farmers' Market.

Pyle, J. (1971). Farmers' markets in the United States: Functional anachronisms. *Geographical Review, 61*(2), 167–197.

Rhoades, R. (2001). Breaking new ground: Linking research, participation, and sustainable development. In R. Rhoades (Ed.), *Bridging human and ecological landscapes, participatory research and sustainable development in an Andean agricultural frontier* (pp. 3–24). Dubuque, IA: Kendall/Hunt Publishing.

Robbins, P. (2004). *Political ecology, a critical introduction.* Malden, MA: Blackwell Publishing.

Salant, P., & Dillman, D. (1994). *How to conduct your own survey.* New York: John Wiley and Sons, Inc.

Saylor, K. (2003). *Evaluation of the 2001 Pilot Oregon Senior Farmers' Market Nutrition Program.* Unpublished master's thesis, Oregon State University, Corvallis, OR.

Scoones, I. (1999). New ecology and social sciences: What prospects for a fruitful engagement? *Annual Review of Anthropology, 28*, 479–507.

Sheridan, T. (1988). *Where the dove calls, the political ecology of a peasant corporate community in northwestern Mexico.* Tucson, AZ: University of Arizona Press.

Sherry, J. (1987). Heresy and the useful miracle: Rethinking anthropology's contribution to marketing. In J. Sheth (Ed.), *Research in marketing* (pp. 285–306). Greenwich, CT: JAI Press, Inc.

Sims, B., & Bentley, J. (2002). Participatory research: A set of tools but not the key to the universe. *Culture and Agriculture, 24*(1), 34–41.

Smith, M. (1960). *An analysis of the East Cleveland Farmers' Market.* Unpublished doctoral dissertation, The Ohio State University, Columbus, OH.

Smith, R. (1972). *Periodic markets in Africa, Asia and Latin America* (Exchange Bibliography, No. 318). Monticello, IL: Council of Planning Librarians.

Smith, R. (1979). Periodic market-places and periodic marketing: Review and prospect I. *Progress in Human Geography, 3*, 471–505.

Smith, R. (1980). Periodic market-places and periodic marketing: Review and prospect II. *Progress in Human Geography, 4*, 1–31.

Smith, S., & Reeves, E. (1989). Introduction. In Smith, S. & Reeves, E. (Eds.), *Human systems ecology, studies in the integration of political economy, adaptation, and socionatural regions* (pp. 1–18). Boulder, CO: Westview Press.

Sommer, R., Wing, M., & Aitkens, S. (1980). Price savings to consumers at farmers' markets. *The Journal of Consumer Affairs, 14*(2), 452–462.

Spitzer, T., & Baum, H. (1995). *Public markets and community revitalization.* Washington, DC: ULI—The Urban Land Institute and Project for Public Spaces.

Steiner, F. (2002). *Human ecology, following nature's lead.* Washington, DC: Island Press.

Stephenson, G., & Lev, L. (1998). *Support for local agriculture in two Oregon cities* (Oregon Small Farms Tech. Rep. No. 1). Corvallis, OR: Oregon State University Extension Service. Retrieved March 10, 2006, from http://smallfarms.oregonstate.edu/publications/techreports/TechReport1.pdf/

Stephenson, G., & Lev, L. (2004). Common support for local agriculture in two contrasting Oregon communities. *Journal of Renewable Agriculture and Food Systems, 19*(4), 210–217.

Stone, G. (2004). Biotechnology and the political ecology of information in India. *Human Organization, 63*(2), 127–140.

Stonich, S. (1995). The environmental quality and social justice implications of shrimp mariculture development in Honduras. *Human Ecology, 23*(2), 143–168.

Stonich, S. (1999). Comments: After nature: Steps to an antiessentialist political ecology by Arturo Escobar. *Current Anthropology, 40*(1), 23–24.

Stonich, S., & DeWalt, B. (1989). The political economy of agricultural growth and rural transformations in Honduras and Mexico. In S. Smith & E. Reeves (Eds.), *Human systems ecology, studies in the integration of political economy, adaptation, and socionatural regions* (pp. 202–230). Boulder, CO: Westview Press.

Sutton, M., & Anderson, E. (2004). *Introduction to cultural ecology.* Walnut Creek, CA: Altamira Press.

Thiemann, G. (1983). *An economic analysis of supplier motivations at farmers markets.* Unpublished master's thesis, Oregon State University, Corvallis, OR.

Thiemann, G., & Burt, L. (1983). *Developing and managing a farmers' market in Oregon* (Extension Circular 1116). Corvallis, OR: Oregon State University Extension Service.

Thilmany, D., & Watson, P. (2004). The increasing role of direct marketing and farmers' markets for western U.S. producers. *Western Economics Forum, 3*(2), 19–25.

Thu, K., & Durrenburger, E (Eds.). (1998). *Pigs, profits and rural communities.* New York: State University of New York Press.

Townsend, P. (2000). *Environmental anthropology, from pigs to policies.* Prospect Heights, IL: Waveland Press.

Trotter, R., & Schensul, J. (1998). Methods in applied anthropology. In H. Bernard (Ed.), *Handbook of methods in cultural anthropology* (pp. 691–735). Walnut Creek, CA: Altamira Press.

Tweeten, L., & Flora, C. (2001). *Vertical coordination of agriculture in farming-dependent areas* (Task Force Rep. No. 137). Ames, IA: Council for Agricultural Science and Technology.

UC SAREP. (n.d.). *What is a sustainable community food system?* Davis, CA: University of California, Sustainable Agriculture Research and Education Program. Retrieved March 5, 2008, http://www.sarep.ucdavis.edu/cdpp/cfsdefinition.htm

USDA. (1998). *A time to act: A report of the USDA National Commission on Small Farms*. Washington, DC: United States Department of Agriculture.

Van de Fliert, E., & Braun, A. (2002). Conceptualizing integrative, farmer participatory research for sustainable agriculture: From opportunities to impact. *Agriculture and Human Values, 19*, 25–38.

Wann, J., Cake, E., Elliot, W., & Burdette, R. (1948). *Farmers' produce markets in the United States. Part 1, History and description* (Marketing Research Rep. No. 17). Washington, DC: U.S. Department of Agriculture, Farm Credit Administration.

Wimberley, R., Reynolds, W., Vander Mey, B., Wells, B., Ejimakor, G., Bailey, C., et al. (2003). *Food from our changing world: The globalization of food and how Americans feel about it.* Raleigh, NC: Department of Sociology, North Carolina State University.

Retrieved March 5, 2008, from http://sasw.chass.ncsu.edu/global-food/foodglobal.pdf

Wright, W., & Middendorf, G. (2008). Fighting over food: Change in the agrifood system. In W. Wright & G. Middendorf (Eds.), *The fight over food: Producers, consumers, and activists challenge the global food system* (pp. 1–26). University Park, PA: The Pennsylvania State University Press.

Wuest, S., McCool, D., Miller, B., & Veseth, R. (1999). Development of more effective conservation farming systems through participatory on-farm research. *American Journal of Alternative Agriculture, 14*, 98–102.

Zimet, D., Hewitt, T., & Henry, G. (1986). Characteristics of successful vegetable farmers' retail markets. In *Proceedings of Florida State Horticulture Society 99* (pp. 291–293). Tallahassee, FL: Florida Agricultural Experiment Station.

Zimmerman, M. (1937). *Super market: Spectacular exponent of mass distribution.* New York: Super Market Publishing.

INDEX

www.ingramcontent.com/pod-product-compliance
Lightning Source LLC
LaVergne TN
LVHW101501200926
848483LV00001B/189

9781604975093